SCIENCE SKILLS
Problems in GCSE Science

J Bennetts

Senior Lecturer in Science Education
St Martin's College
Lancaster

P Burdett

Head of Science
Thornes House High School
Wakefield

J Mundie

Lecturer in Science and Engineering
East Warwickshire College
Rugby

A Williams

formerly
Head of Science
Benfield School
Newcastle-upon-Tyne

HODDER AND STOUGHTON
LONDON SYDNEY AUCKLAND TORONTO

Contents

Note

The following symbols appear beside some questions in the text. See Introduction for details:

* Questions which require some knowledge of facts, concepts and principles.

A Questions for which a discussion of the answer is provided

Introduction: mainly for teachers

The DES National Criteria for Science

These criteria require that about 45% of GCSE assessment be recall and understanding of syllabus content: facts, theories and principles. At least 20% but usually not more than 30% is based on coursework assessment of practical skills. Thus between 25% and 35% of the marks will be for the process skills of handling information and solving problems.

The National Criteria document lists sixteen skills and abilities. (3.2.1–3.2.16). Some of these are listed below and we have grouped them under four main headings.

Information handling
This includes:

(a) translating information from one form to another,
(b) extracting from available information data relevant to a particular context.

Problem solving
This involves applying scientific ideas and methods to solve quantitative and qualitative problems.

Patterns
Students are expected to recognise patterns in data, form hypotheses and deduce relationships.

Evaluation
This includes:

(a) drawing conclusions from data and evaluating it critically,
(b) recognizing and explaining variability and unreliability in experimental measurement.

Other skills in the National Criteria for Science are:

(a) Communicating scientific ideas and arguments logically, concisely and in various forms;
(b) suggesting scientific explanations of unfamiliar facts, observations and phenomena;
(c) making decisions based on examination of evidence and arguments;
(d) recognizing that the study and practice of science are subject to various limitations and uncertainties;
(e) explaining technological applications of science and evaluating associated social, economic and environmental implications.

These skills cannot be placed under any one heading, they should permeate an entire assessment. Indeed the final skill in the list is to be assigned a weighting of not less than 15%.

It is important to realize that the National Criteria for Science have an overarching status with regard to biology, chemistry and physics. Syllabuses and examinations in these subjects must meet the same criteria.

A more recent document on Grade Criteria for Science (SEC, June 1986), has three domains: A knowledge, B handling information and solving problems and C practical enquiry. It is suggested that domain B should have a 50% weighting. There are nine subdomains. Again, we have grouped these under four main headings.

Information handling
B1 Locate, select or organize information from a variety of sources.
B2 Translate information from one form to another.

Problem solving
B7 Solve quantitative problems.
B9 Prepare solutions to a given design problem.

Patterns
B3 Recognize or describe patterns in information.
B4 Use patterns to make predictions.

Evaluation
B6 Evaluate statements or arguments.
B8 Criticize or suggest improvements to presented experimental procedures.

Interwoven with these there is
B5 Explain novel phenomena or relationships by hypothesis or application of knowledge.

The Draft Grade Criteria for Science are matched by parallel documents for biology, chemistry and physics.

There are more than ten Mode 1 double subject syllabuses in science. Many of these can also be used for single certification. In addition there are a number of single subject syllabuses in science. Besides this there are many Mode 3 schemes, some of which have a candidature exceeding 1000.

One examining group has chosen to print in its syllabus the sixteen skills and abilities exactly as given in the National Criteria. Other examining groups have chosen to cluster them under main headings such as application, analysis, synthesis and evaluation.

Two of our main headings are *Information handling* and *Problem solving*. Together, they give a clear picture of the nature of the processes and skills in science.

Patterns: seeking and using patterns, could be considered to be an extension of information handling. *Evaluation* is often an appraisal of solutions to problems. It can also be a commentary on the way data have been interpreted or the way patterns have been used to make predictions.

Longer questions inevitably proceed from one activity to another. Questions which begin with information handling could develop into pattern seeking and evaluation. We have therefore used a fifth heading, *Synthesis*, to accommodate this kind of material.

In some instances, candidates will be required to exercise these skills without the need to remember facts, theories and principles from the syllabus. Necessary data for novel situations will be presented. Some would say that this is the only fair way to assess the skills. Such questions are described as 'content free'. To reach the answer, recall is incidental.

Inevitably, many assessment items which test science skills must be 'content free'. Remembering facts is not very important. If scientists want to know the melting point of aluminium or the boiling point of ethanol, they can look up these facts. The ability to manipulate the data is the essence of science.

Most of our questions are 'content free'. Those which require some knowledge of facts, concepts and principles are marked with an asterisk *.

National Criteria for biology, chemistry and physics contain essential content. This is the 'National Core' for each subject and is intended to represent about two-thirds of the content of any syllabus. At the time of writing, there is no stipulated content for science. The content of most double subject syllabuses approximates to the sum of the National Cores for biology, chemistry and physics, though this of course is not necessarily the case for some particular selections within modular schemes. Where questions in this book assume syllabus content, this is restricted to that which appears in the National Core for the three subjects.

Where a chapter contains *multiple choice questions*, these are grouped at the beginning, with easy questions first and harder ones later. Throughout, questions are arranged to provide an incline of difficulty.

GCSE requires that examinations provide opportunities for all candidates to show what they know, understand and can do. In biology, chemistry and physics, National Criteria have been interpreted to mean about 50% to 70% common questions and the remainder on components targeted towards lower or upper ability levels. Some science syllabuses use this model. In papers targeted at grades G to C, it is expected that the F level candidate will be able to do about 40% of the material. This means that there will be some questions which are easier than the type of questions usually seen on CSE papers in the past. Some syllabuses have papers targeted at grades B and A only, whereas GCE 'O' level aimed at grades E to A, so these papers are likely to contain questions harder than anything ever seen at GCE 'O' level. In order to facilitate positive achievement at both ends of the ability spectrum, GCSE has extended the range of difficulty in questions designed for the sixteen-year-old candidate. It will be noticed that questions towards the beginning of a chapter seem 'very easy' while those towards the end seem 'very hard'.

In science, some syllabuses do not involve differentiated components. It is maintained that candidates can make a response at a lower, intermediate or higher level, all experiencing positive achievement. Some of the questions in this book permit responses at different levels.

Where questions break down into a number of subsections, marks for each part have been given. These are intended to indicate possible relative weightings within a question, for example, drawing graph (5), two readings from graph (1 + 1), deduction with reason (2).

The relative weighting given to entire questions would of course depend on whether they were on papers targeted at lower or higher grades. Easy questions would have a low weighting on a high-level paper and by the same token, hard questions would have a low weighting on a low-level paper. By similar reasoning, the harder sections of a particular question do not necessarily get more marks than the easier sections. It depends on the ability range for which the question is intended. This has always been an important principle in setting examination questions. That is why in the paragraph above we claim to show only possible relative weightings within questions.

A A number of questions are flagged with this letter. In the final chapter the answering of these questions is discussed. In most cases, these are not simply mark schemes, but rather an analysis of what the question was designed to test and advice on how to approach it.

It has already been stated that technological applications and social and environmental implications must feature prominently in GCSE Science. We hope that the collection of questions in this book reflects this admirable requirement.

Finally, it is neither possible nor proper to separate the skills and processes of science into neat compartments. The collation of questions under each heading has been done on a 'best fit' basis. We freely admit that in *Information handling* some questions contain elements of problem solving, patterns and evaluating. Perhaps the best questions are those that cannot be categorized.

1 Information handling

A

1

Mr Khan has a factory which makes throw-away plates for hot meals. He has been offered a material called Foodiflex. Some properties of Foodiflex are listed below. Which one of these should be his main reason for rejecting Foodiflex for making plates?

A. It becomes flexible at 100°C.
B. It is difficult to colour.
C. It tends to crack when hit by a spoon.
D. It is not biodegradable.
E. It is easy to shape.

2

Pieces of apple usually go brown when they are left in air for a few minutes. In food-processing factories it is important to stop this browning. Therefore, cut apples are put in special solutions. The graph shows how browning is controlled by the pH of a solution. Which one of these solutions stops browning?

A. Sodium hydrogen carbonate (pH 9).
B. Sugar (pH 6.5).
C. Citric acid (pH 4).
D. Sodium hydrogen sulphate (pH 2.5).

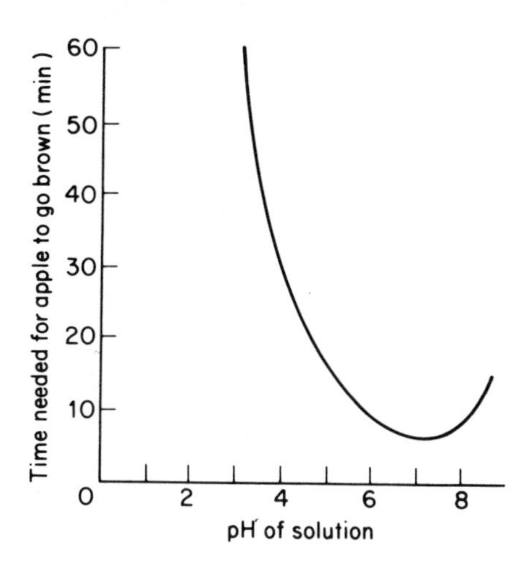

3

The diagram shows what four test-tubes look like when two different chemicals have been mixed together.

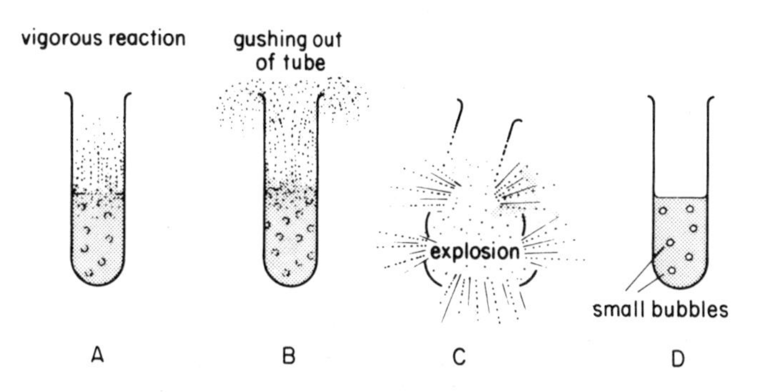

Some chemicals react by changing colour or by giving off a smell, or even by fizzing. Some react more vigorously. Put them in order of how reactive they are with the strongest one first.

4

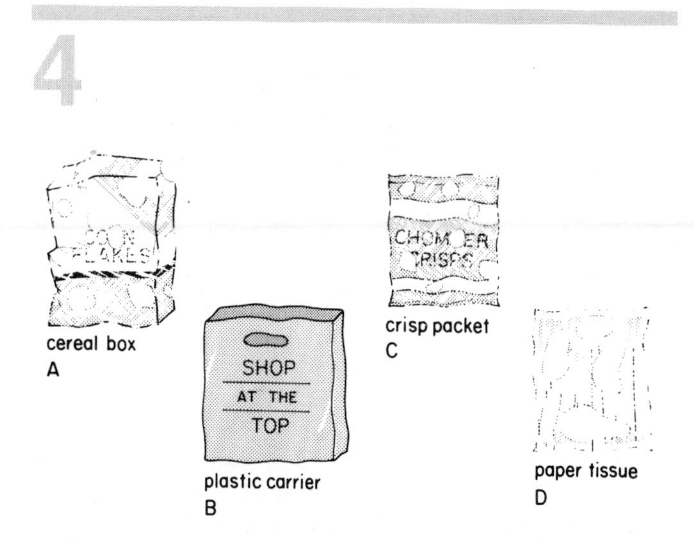

Some materials can be broken down by microbes in the soil and air. They are decomposed. Look at the diagrams of four materials that were left in the soil for three months. Put them in order of decomposition with the most decomposed first.

5

An environmental health inspector checks milk using a blue dye called Resazurin. If there are no bacteria, the dye in the milk will stay the same colour when the milk is kept at 37°C. If there are a large number of bacteria in the milk, the dye changes colour. It goes to purple, then pink and then white.

The inspector tested milk from four dairies, A, B, C and D. The samples of milk were all kept at 37°C in the same conditions. Her results are shown below.

Sample	*Colour after 10 min*	*Colour after 1 hour*
A	blue	white
B	blue	pink
C	pink	white
D	blue	blue

(a) Which of the samples contains most bacteria?
(b) Which of the samples contains least bacteria?

6

Three-pin plugs need a cartridge fuse. A cartridge fuse contains a length of wire. The safe working current and the current which will make the fuse blow depend on the diameter of the wire.

Diameter (mm)	*Safe working current* (ampere)	*Fusing current* (ampere)
0.50	2.0	3.0
0.55	2.3	3.5
0.60	2.6	4.0
0.65	3.3	5.0

A 3 A fuse should allow 3 A to pass but not much more. Which wire should be used by a workshop for making cartridge fuses which are rated at 3 A?

A. 0.50 mm diameter.
B. 0.55 mm diameter.
C. 0.60 mm diameter.
D. 0.65 mm diameter.

7

The diagrams show how four paper bridges were tested for strength. Put them in order of strength with the strongest one first.

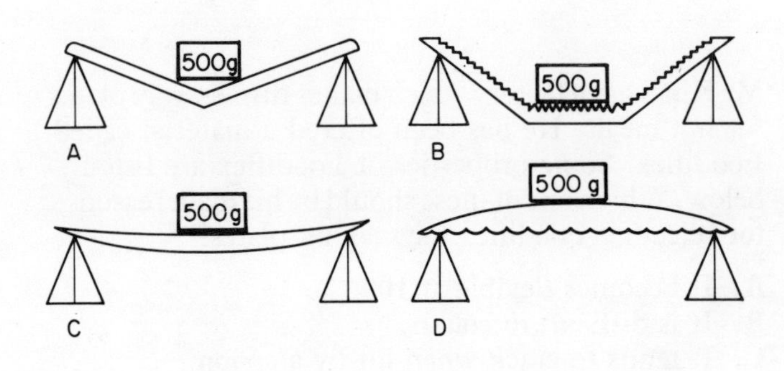

*8

These pie charts show the proportions of clay, silt and sand in four different soils. Use the diagrams to answer the questions below.

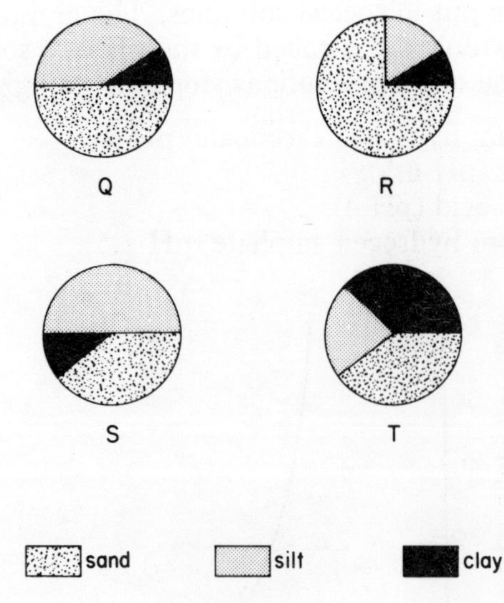

(a) Which soil will have the slowest drainage rate:
(i) Soil Q,
(ii) Soil R,
(iii) Soil S,
(iv) Soil T?

(b) Which soil is likely to need extra minerals in the form of fertilizer:
   (i) Soil Q,
   (ii) Soil R,
   (iii) Soil S,
   (iv) Soil T?

(c) Which soil should produce the best crops:
   (i) Soil Q,
   (ii) Soil R,
   (iii) Soil S,
   (iv) Soil T?

9

One way of testing for fat is to rub the food sample on to paper. The paper lets light through a greasy spot. Mr Lule tested a number of foods eaten by people on the island of Pemba.

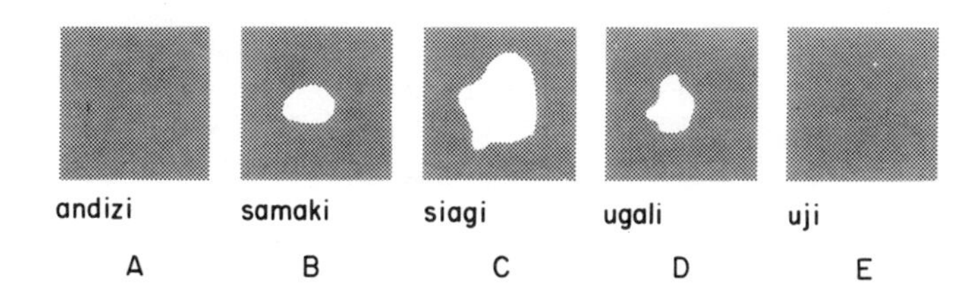

(a) Which ones had no fat? (*1*)
(b) Which one had the most fat? (*1*)

*10

A B C

plaque

The diagram shows three sets of upper teeth. Disclosing tablets have been used to show up plaque. These people had just cleaned their teeth, but not equally well. Put them in order with the worst cleaned teeth first.

11

Fabric	Types of dyes					
	A	B	C	D	E	F
Acetate	very fast	very fast	will not dye	will not dye	will not dye	very fast
Nylon	very fast	very fast	medium fast	medium fast	will not dye	medium fast
Acrylic	will not dye	will not dye	will not dye	will not dye	will not dye	will not dye
Cotton	medium fast	very fast	medium fast	medium fast	medium fast	will not dye
Polyester	medium fast	medium fast	medium fast	very fast	medium fast	will not dye
Wool	medium fast	very fast	medium fast	very fast	will not dye	very fast
Silk	will not dye	very fast	very fast	very fast	will not dye	will not dye
Linen	will not dye	very fast	medium fast	medium fast	very fast	will not dye

■ = very fast
◪ = medium fast
□ = will not dye

The chart shows how fast (resistant to running) dyes are with certain fabrics. This was produced as a guide for people wishing to dye their own clothes.

(a) Which fabric cannot be dyed at at all? (*1*)
(b) Which dye can dye most fabrics? (*1*)
(c) Which fabric can only be dyed very fast in dyes B, D and F? (*1*)

12

Study the pie chart. It shows what makes up crystal glass. (Each segment is worth 10%)

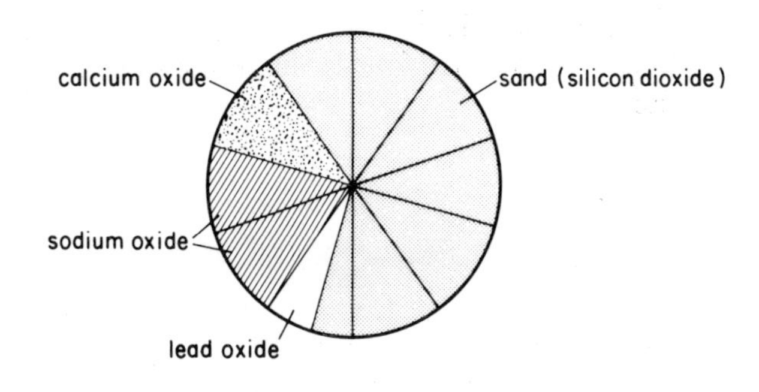

(a) What percentage of it is calcium oxide? (*1*)
(b) What percentage of it is lead oxide? (*1*)
(c) Which substance is more than the rest put together? (*1*)

13

Clothes are made from many fabrics. People were asked to look through their clothes and find out the types of fabric. The results are given as percentages.

Fabric type	*Percentage* (%)
Polyester	25
Wool	20
Silk	5
Cotton	40
Nylon	10

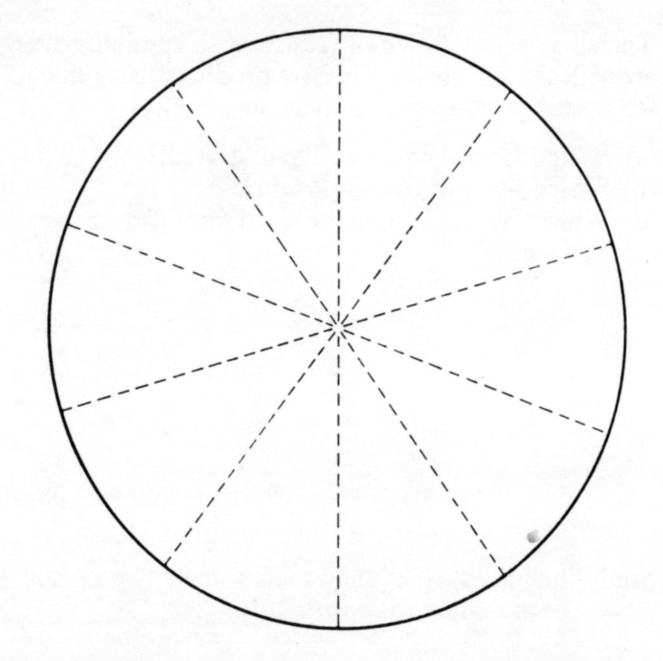

Copy the pie chart. Each segment is worth 10%. Use the figures above to complete it. Shade and label it or give a key.

14

Look at the pictogram. It shows the number of litres of chlorine produced from the same quantity of six different bleaches. The more chlorine the better the bleach.

(a) How many litres of chlorine were produced by Bluloo? *(1)*
(b) Which bleach produced most chlorine? *(1)*
(c) Which bleach is only half as good as Biodot? *(1)*

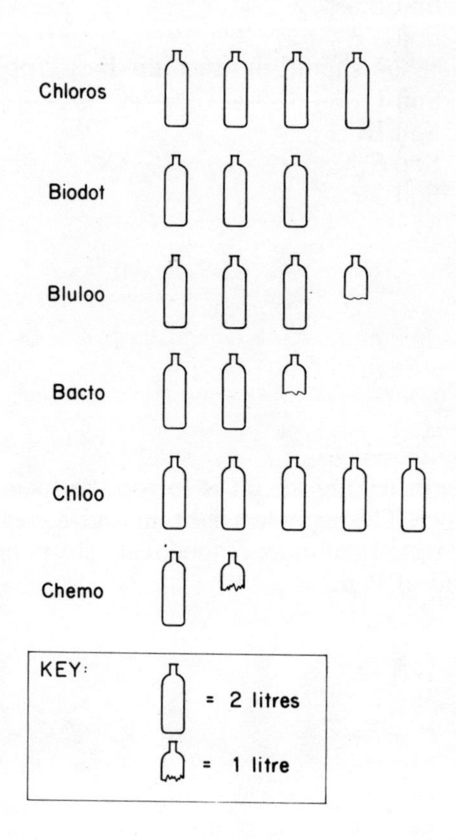

15

Here is a bar chart showing the accidents that happened in some science labs over one year.

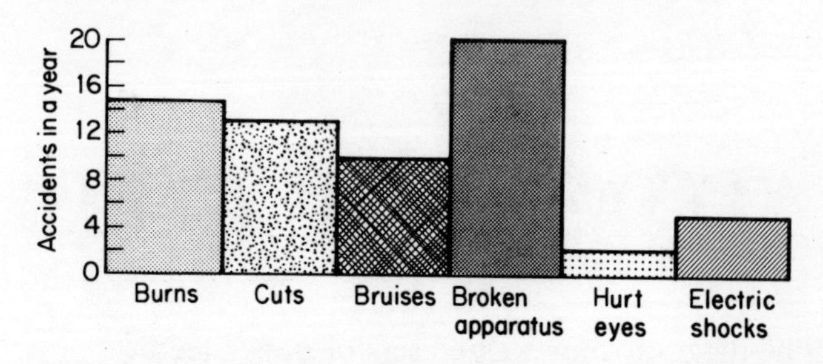

(a) Which accident was the commonest? *(1)*
(b) Which accident was the rarest? *(1)*
(c) Which two types were less common than bruising? *(2)*

*16

The table here shows the masses (in grams) of male and female animals, including humans.

Mass (g)	600	700	400	500	200	300	20 000	20 000
Animal	rat	rat	gerbil	gerbil	mouse	mouse	young human	young human
Sex	female	male	female	male	female	male	female	male

(a) (i) Which are generally heavier, males or females? (*1*)
(ii) Suggest why this occurs. (*1*)
(b) Which animals have the same mass for males and females? (*1*)
(c) Which animal has the lowest mass? (*1*)
(d) At what stage in human development do the masses of males and females begin to show differences? (*1*)

17

Some pupils were interested in which bones were most likely to be broken. They asked 35 first years who had had broken bones. The results were:

Bone	Forearm	toe	wrist	thigh	shin	collar	rib	finger
No. of people	12	2	1	4	3	5	6	4

(a) Copy and complete this bar chart. (*3*)

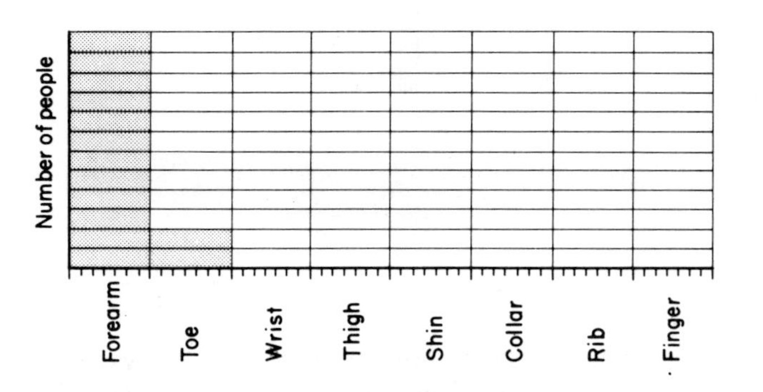

(b) Which bone was broken most often? (*1*)
(c) Which bone listed above was broken least? (*1*)
(d) What was the total number of bones broken? (*1*)
(e) The number of pupils was only 35 and yet the total number was more. How do you explain this? (*1*)
(f) Do you think that these results fairly represent those for the whole school? Give a reason. (*2*)

18

Mrs Bhati is a health and safety inspector. She needs to check that workers have enough light to do their job. Mrs Bhati measures the light intensity in W/m^2 along a workbench. The graph below shows how intensity changes with distance from end A.

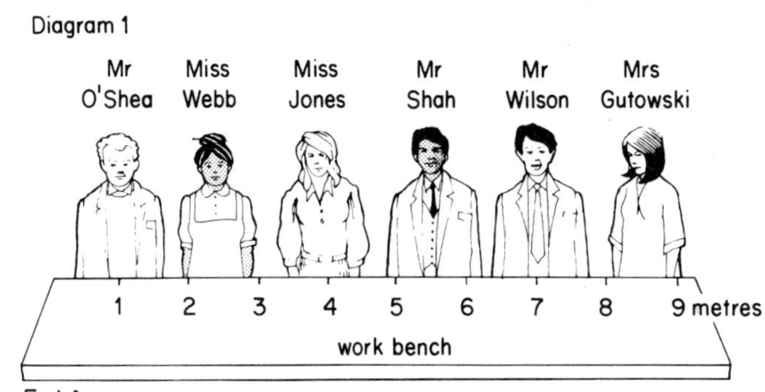

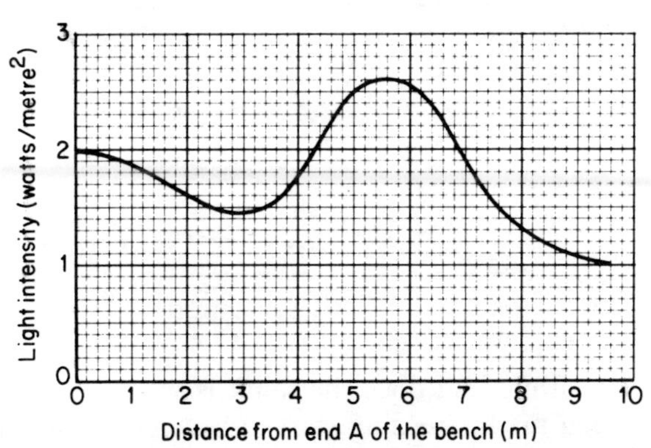

(a) One end of the workbench is near a window. Which end is it?
(b) The rules for this factory say that for each worker the light intensity should be not less than 1.7 W/m^2.
(i) Which workers need more light?
(ii) Where should extra lights be placed?

A

19

Study the diagram of the apparatus. It is a device for measuring breathing rate. Explain how it works and what X and Y represent.

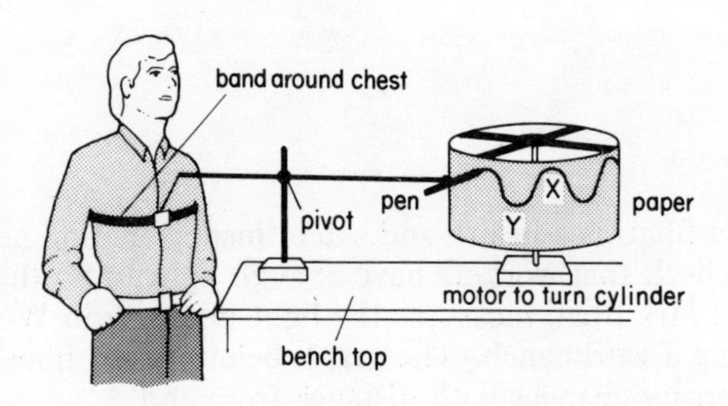

20

A food scientist produced the graphs here to show how long she thought food could be preserved using different methods.

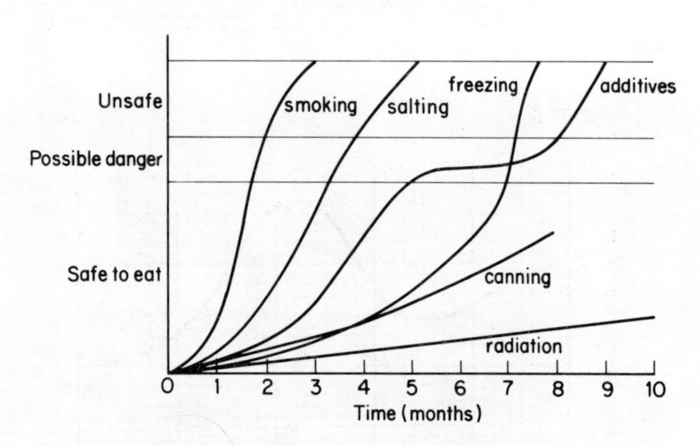

(a) Which method does she think is worst? (*1*)
(b) When might freezing start to become a possible danger? (*1*)
(c) One method has a different curve from the rest. Which one? How long is its possible danger period in months? (*2*)
(d) Which method keeps food longest? (*1*)
(e) The canning line goes to 8 months. After how many months do you think canned food would become possibly dangerous? (*1*)

*21

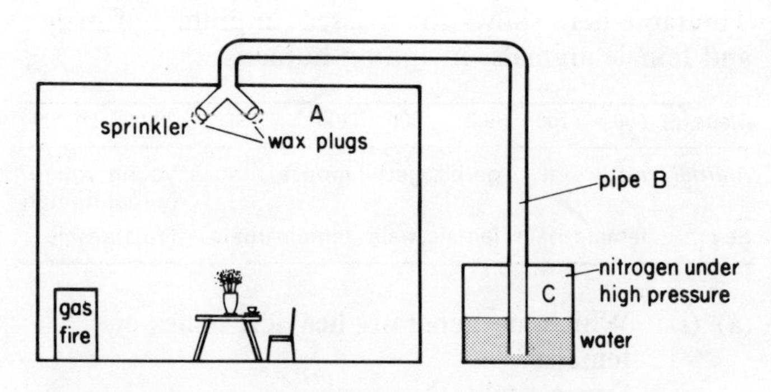

The diagram shows one type of automatic fire extinguisher. Explain simply how it might work. Mention parts (A), (B) and (C).

22

Look at the diagram of a new type of aerosol container. These aerosols are used to spray a variety of liquids. Describe simply how it might work.

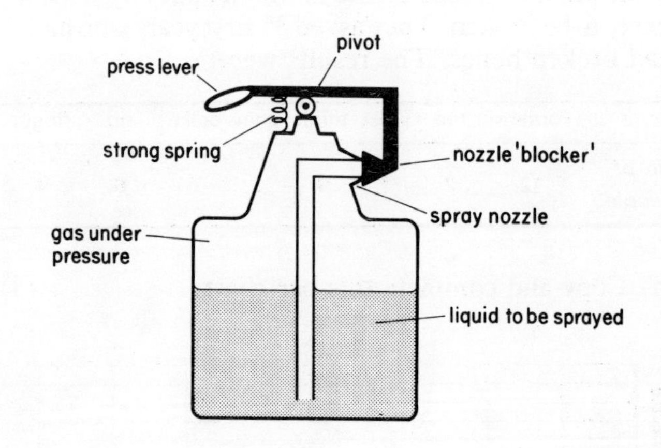

*23

The diagram at the top of the next page shows some electrical equipment connected together to form a circuit. Draw a circuit diagram using appropriate symbols for each component.

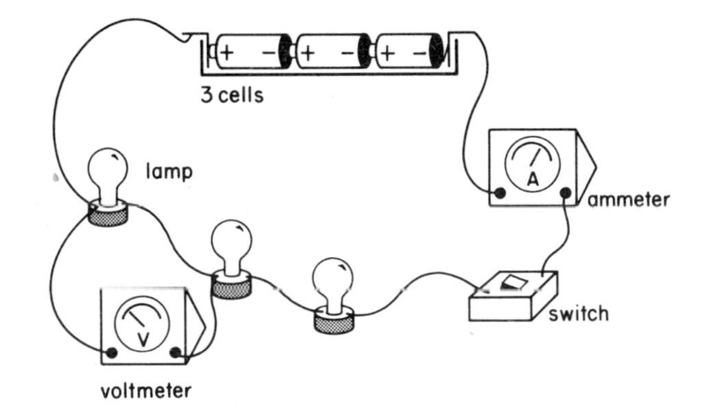

24

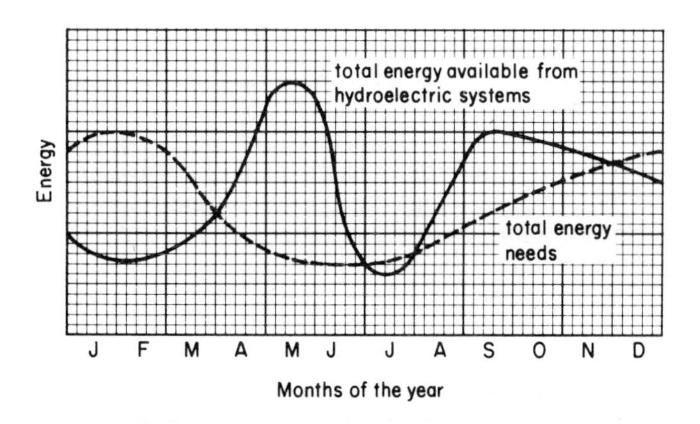

The graph shows the electrical energy used by a country during twelve months of the year. It also shows the electrical energy which is available from hydroelectric systems in the country. During which months did the country have to produce some electrical energy in other ways?

25

When a photograph is taken, an image is formed on the film. At first the image is invisible. Developing solution changes the invisible image into one which we can see. It takes a few minutes to work. Temperature affects developing time. Here are some data showing the developing time needed for some black and white film at various temperatures.

Temp (°C)	16	17	18	19	20	21
Time (min)	10	8	6.5	5.5	5	4.5

(a) Plot these data on a graph. Put temperature on the upright axis and time along the bottom. Label the axes. Join up the points to give the best smooth line. *(5)*

(b) What developing time is needed for a temperature of 17.5°C? *(1)*

(c) At what temperature would a film develop in 9 min? *(1)*

(d) What happens to the developing time as the temperature increases? *(1)*

26

Diagram 1 shows how the amount of light and the amount of carbon dioxide change with depth in a lake. The wider the column, the greater the amount.

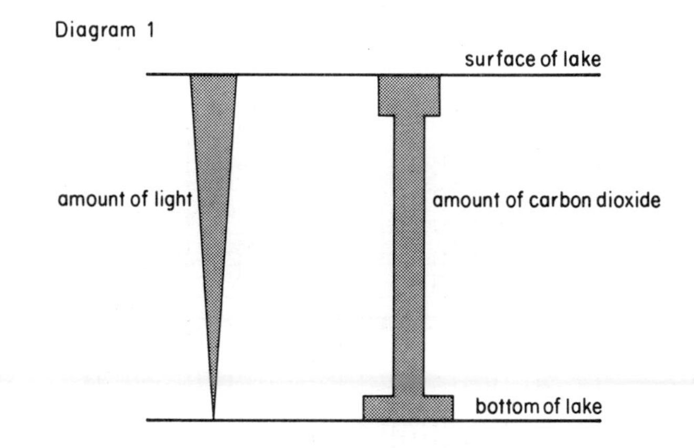

Certain plants grow in the lake and like all plants they need to photosynthesize.

Things needed for photosynthesis	*Things produced by photosynthesis*
carbon dioxide	oxygen
water	carbohydrate
light	

Use this information to decide where the plants are most likely to grow in the lake. Diagram 2 shows the lake.

Diagram 2

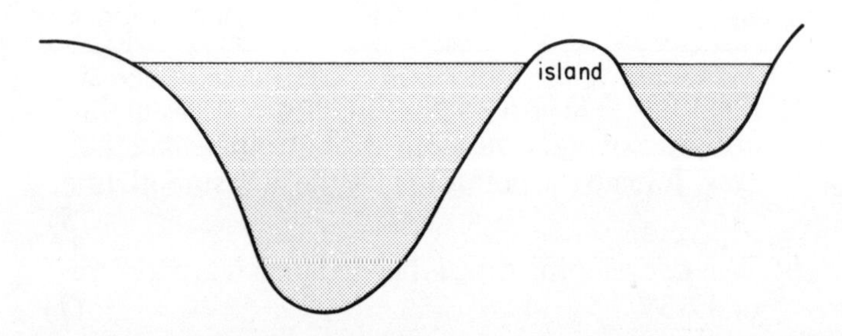

(a) Copy diagram 2. Mark with a P all the places where water plants might grow. *(6)*
(b) Explain how you decided where plants will grow. *(4)*

27

The graph shows how the e.m.f. (voltage) of a car battery depends on the concentration of the acid in it.

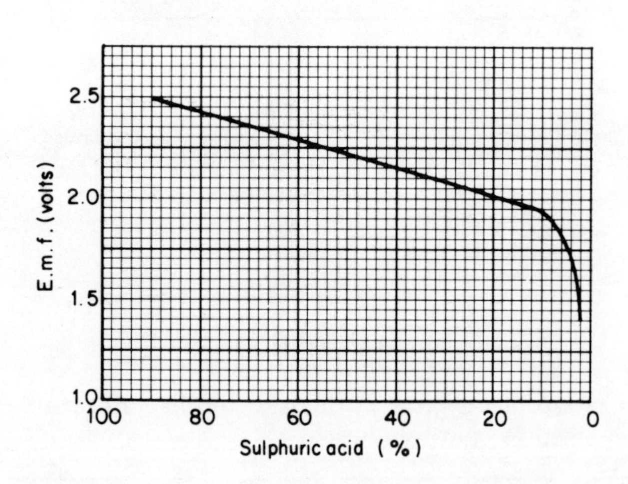

(a) What is the e.m.f. of the battery when the acid has a concentration of 5%? *(1)*
(b) In Tanzania, garages in small villages do not have electricity. This can be a problem for motorists living in villages who find that their car battery is 'flat'. But garages sometimes have a bottle of concentrated acid. What increase in e.m.f. would you expect if acid concentration is changed from 5% to 30%? *(2)*

28

A chemical factory has some glassware which has not been used for many years. Most of it contains dirty solids. The glassware is labelled 'Toxicity Unknown'. This means that nobody knows what the dirty solids are and they may be harmful. A technician decides to use a flame test to find out what the solids might be.

A wire is dipped into each solid and the wire is held in a bunsen flame. When certain metals are present in a substance the flame goes an unusual colour.

Metal	*Colour*
potassium	magenta (lilac)
lead	blue
calcium	red
sodium	yellow
copper	green

Sometimes there are at least two metals present in a substance. So the bunsen flame could be showing yellow light and blue light at the same time. Here is some information about what happens when colours of light are added together.

green	+	blue	= turquoise
blue	+	red	= magenta
red	+	green	= yellow

(a) The technician finds that the dirty solid in all flasks gives a blue flame. What could be in the flasks? *(1)*
(b) He finds that the dirty solid in all beakers gives a magenta flame. What could be in the beakers? *(3)*
(c) He finds that the dirty solid in all test-tubes gives a yellow flame. What could be in the test-tubes? *(3)*

29

Clearfilm is used to cover dishes, plates, bowls etc. It is made of a plastic that can stretch. Some clearfilm was tested by a group of students to see how it stretched when pulled. The pulling force is measured in newtons (N). It was tested until it broke. The graph at the top of the next page shows the results.

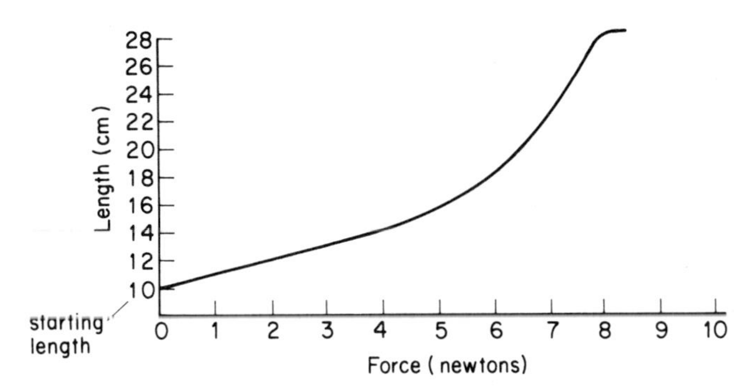

(a) What was the clearfilm's starting length? (*1*)
(b) What was its length when it broke? (*1*)
(c) What force broke it? (*1*)
(d) How far did it stretch up to 5 N? (*2*)
(e) It stretched very differently after this force of 5 N. Describe the way it stretched with loads of more than 5 N. In what way is this stretching different from the stretching for loads of less than 5 N? (*2*)

30

Study the table below.

Food (per 100 g)	*Nutrients*				
	Fibre (%)	*Energy* (kJ)	*Protein* (g)	*Vit A* (μg)	*Vit C* (mg)
cod in batter	0	830	20	0	0
eggs	12	610	12	140	0
raw cabbage	35	90	4	50	50
sausage	0	1500	10	0	0
chips	0	1000	5	0	20
milk	0	270	3	40	1

Everybody needs all of the nutrients in their diet.

(a) What would a person who ate fish, chips, sausage and milk lack in their diet? (*1*)
(b) What is the total energy in 100 g sausage + 200 g cabbage + 100 g milk? (*2*)

31

A test strip was made by some pupils. This was to see how long some photographic papers needed to be exposed to give good photos. If the paper is exposed for too long, it goes black. Here are the results.

Seconds	*Paper* P	*Paper* Q	*Paper* R
1	white	white	grey
3	light grey	white	dark grey/black
5	grey	light grey	very black
7	dark grey/black	grey	all black
9	very black	dark grey/black	all black
11	all black	very black	all black

(a) What time seems best for paper Q? (*1*)
(b) Which paper is the most sensitive? (*1*)

32

Below is a pie chart showing some of the causes of back problems in children and adults.

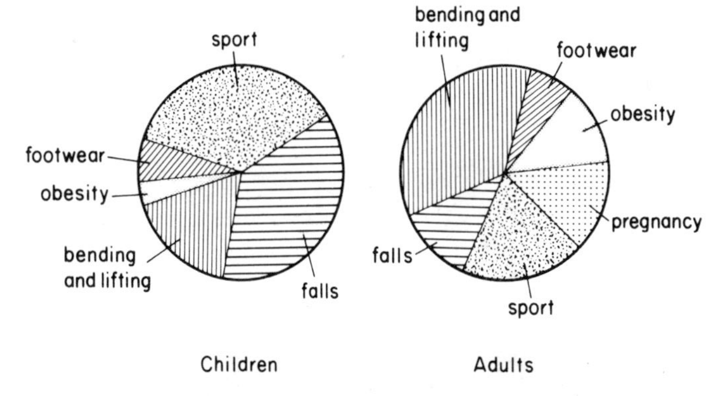

(a) What is the commonest cause in children? (*1*)
(b) What is the least common cause in adults? (*1*)
(c) What cause is equally common in children and adults? (*1*)
(d) Suggest reasons why the major causes are different in adults and children. (*2*)

33

Some pupils wanted to test indicators made from plants. They mixed together juices extracted from four or five plants. They added drops of the mixture to liquids of different pH. These are the colours they got:

pH	1	3	5	7	9	11	13
Colour	pink	blue	red	purple	yellow	yellow	green

Some unknown new liquids were tested with their indicator mixture.

This is what they found:

Liquid	A	B	C	D	E	F
Colour	green	red	yellow	pink	blue	purple

Use the tables to answer these questions:

(a) What is the pH of the following: B, D and E? *(3)*
(b) Which liquid has a pH of 13? *(1)*
(c) It is not possible to find out accurately the pH of one of the liquids. Say which one and give a reason for your answer. *(2)*

34 Balancing breakfast

The chart shows amounts of different nutrients in 100 g of four kinds of breakfast cereal.

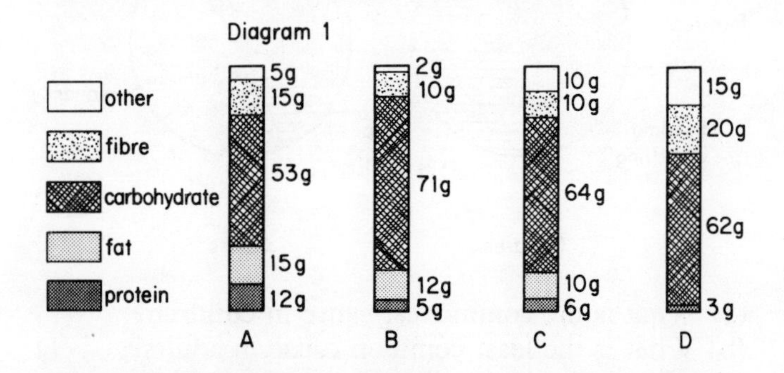

(a) Which of these cereals
  (i) contains most protein, *(1)*
  (ii) contains most fibre, *(1)*
  (iii) contains least fat? *(1)*
(b) This chart shows the amounts of different nutrients in 100 g of milk.

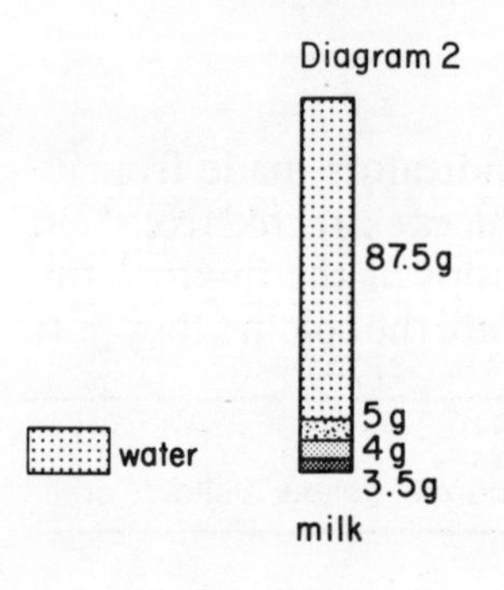

50 g of each cereal is placed in a dish and covered with 100 g of milk. The dishes are labelled A, B, C and D.
  (i) In which dish will the protein be mostly in the cereal? *(1)*
  (ii) Which dish will contain exactly 10 g of fat? *(1)*

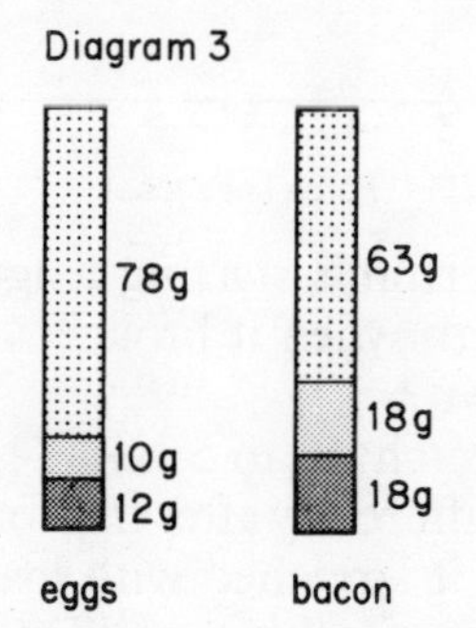

These charts show the amounts of different nutrients in 100 g of bacon and 100 g of egg.
  (iii) Which of the four dishes of milk and cereal should be followed by bacon and eggs to obtain a balanced meal? *(1)*

35 Getting the energy

The main sources of energy in Britain are coal, natural gas and oil. Smaller amounts come from hydroelectric and nuclear power stations. The amount of energy from each source can be given in tonnes of coal equivalent. The table below gives the million tonnes of coal equivalent for each source used in 1960, 1970 and 1980.

Source	*Year*		
	1960	1970	1980
natural gas	0	20	70
oil	68	150	121
hydroelectricity	2	2	2
coal	197	157	121
nuclear	0	10	13

(a) Copy the graph at the top of the next page on to graph paper and complete it. *(5)*

(b) (i) Which source of energy was used less in 1980 than in 1960? *(1)*
  (ii) Suggest a reason for this decline. *(1)*

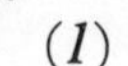

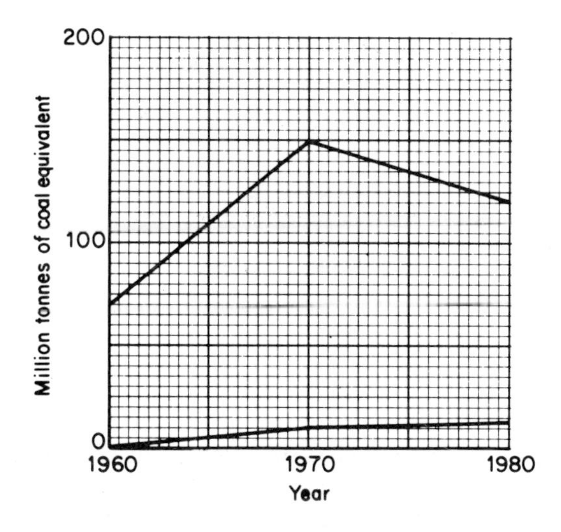

(c) (i) Name the energy source which was used much more in 1980 than in 1970. *(1)*
(ii) Give a reason for its increase in popularity. *(1)*
(d) (i) The use of one energy source has remained constant. Which is it? *(1)*
(ii) Why has there been no change? *(1)*
(e) (i) How much energy in million tonnes of coal equivalent was used altogether in 1960? *(1)*
(ii) How much energy in million tonnes of coal equivalent was used altogether in 1980? *(1)*
(iii) By how much has the total energy used changed? *(1)*
(iv) Give a possible reason for the above change. *(1)*

36

This graph shows the number of hours of sunshine on an average winter day in the centre of London and at Richmond (about 10 miles from the centre).

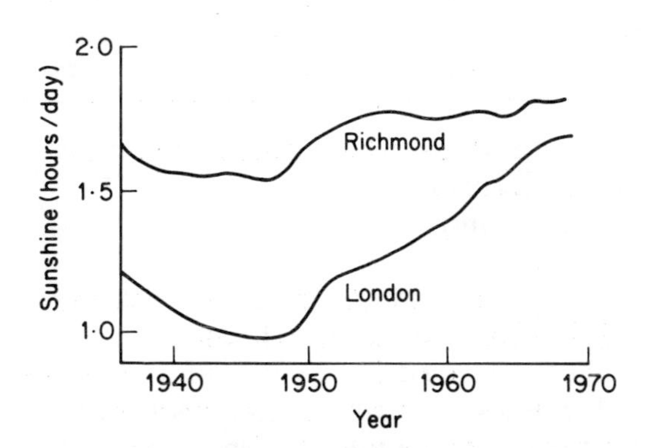

What information can you get from this graph?

*37 Fermentation rates

The experiment was set up as shown. The number of bubbles per minute were counted every 8 hours for three days.

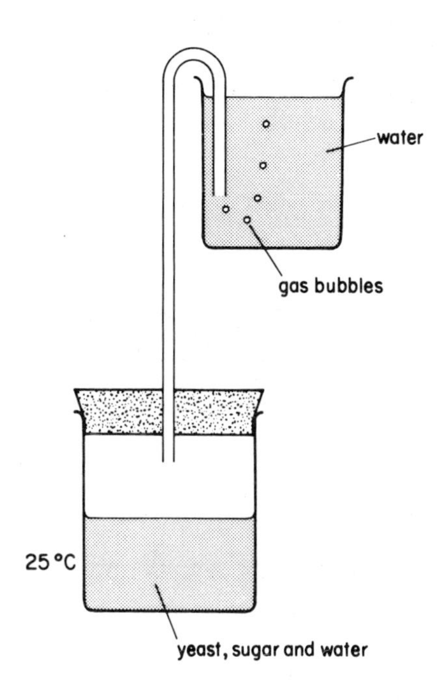

The results are shown in the table.

Time (hr)	start (0)	8	16	24	32	40	48	56	64	72
No. bubbles per min	5	10	25	50	80	80	80	70	30	0

(a) Plot these data on a graph. Put no. of bubbles per minute on the upright axis and time along the bottom. *(5)*
(b) Describe how the rate of bubble production changes over the three days. *(3)*
(c) Suggest two reasons why the change occurs between 48 and 72 hours. *(2)*
(d) Draw your own line on the graph to show what you might expect if you repeated the experiment at 35°C instead of 25°C. Start at the same point. (Your line may not finish at the same point. Note that the same amount of sugar, yeast and water are present.) Label your line 35°C. *(3)*
(e) What is the gas given off? *(1)*

38

A group of pupils wanted to extract dye from red cabbage. Some dyes are temperature sensitive. This means that the colour breaks down if it is heated too much.

These pupils put red cabbage into water at different temperatures. They were trying to find out which temperature was the best for extracting the dye. The graph shows their results.

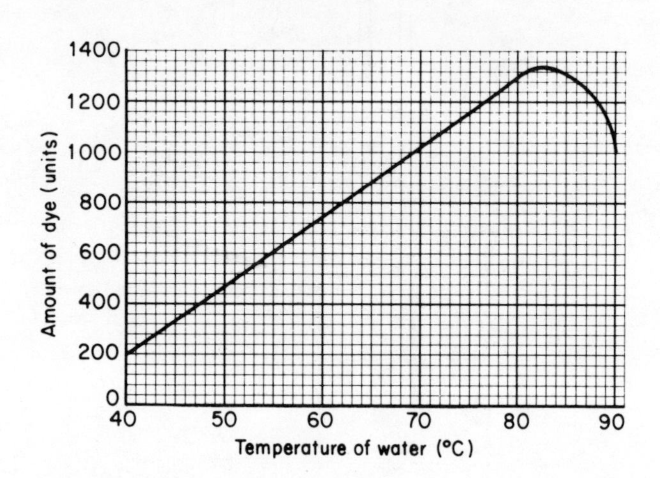

(a) How much dye was extracted at 65°C? *(1)*
(b) What temperature released 500 units of dye? *(1)*
(c) What temperature released most dye? *(1)*
(d) Why is there a sudden change in direction between 80°C and 90°C? *(1)*
(e) The line only goes to 90°C. How much dye do you think would be extracted at 100°C? *(2)*

39

Ship canals need to be inspected. There must be no obstacles at the bottom of the canal which could damage ships. An inspection boat sends pulses of ultrasound to the bottom of the canal every 9/100 of a second. The ultrasound is reflected back. This is an echo effect. An oscilloscope on the ship shows the pulses of ultrasound which are sent out as strong signals.

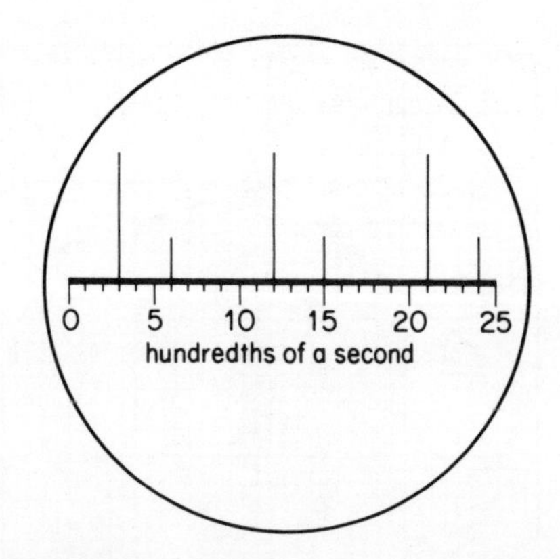

(a) How much time passes between a pulse being sent out and its echo being received? *(1)*
(b) How long does it take for sound to get to the bottom of the canal? *(1)*
(c) How would the picture on the oscilloscope change if the boat passed over a lorry which had been dumped in the canal? *(1)*

*40 Protecting metal

Circulating water in central heating systems contains a substance called an inhibitor. The inhibitor helps to reduce the corrosion inside the radiator. Several inhibitors were tested for a year to see how effective they were, and the results are given in the table.

Inhibitor	*pH*	*Mass of radiator metal lost (mg)*	
		Aluminium	*Iron*
car antifreeze	10	126	197
SNPS	11	0	1293
SBN	8	309	5
ethylene glycol	8	0	5562
control	5–8	436	5172

(a) Select an inhibitor which would be best for radiators made of:
(i) aluminium,
(ii) iron. *(2)*
(b) If you did not know the material of the radiator which inhibitor would you choose? *(1)*
(c) The antifreeze used in cars is mainly ethylene glycol, but also contains small amounts of other substances. Use the information in the table to explain why these substances are added. *(1)*

(d) The control was tap water to which nothing had been added. Why was a control experiment performed? (*1*)
(e) Suggest a reason why iron corrodes more than aluminium. For which inhibitor is this not true? (*2*)
(f) Give a possible reason why all the inhibitors have a pH greater than 7. (*2*)

*41 Moulds as foods

We rely on moulds for the commercial production of many foods. Yeast is a mould. We use one strain of yeast for breadmaking and another for brewing beer. The yeast that ferments wine lives naturally on the skin of grapes. Other moulds are used in cheese making. Cheese makers keep specially pure cultures of moulds and they use them to infect the separated curds. Different moulds make different flavoured cheeses. Lancashire blue vein and Danish blue cheese are infected with strains of blue mould. They live in the cheese, growing in the spaces between the curds. Sometimes the cheese is sold after a few weeks or months, but other blue cheeses like Stilton may be much older when they are sold.

Recently, experiments to use waste food material to grow moulds have given good results. Some sweet factories use mould to recycle sugar waste. Instead of throwing away water containing waste sugars from sweet making, the sweet watery waste is drained into tanks and a mould grows on it. When the mould is ready, it is separated from the waste and processed to give a textured protein that can be turned into animal food. In these factories, moulds are used to recycle food waste and to produce a substance that can be sold to cover expenses.

(a) Name two different foods produced with the help of moulds. (*2*)
(b) When fruit juice or cereals are fermented to produce drinks, what substance is produced by the action of yeast on them? (*2*)
(c) How do cheese makers ensure that their cheeses always taste the same? (*2*)
(d) Milk curds always taste the same. How can cheese makers produce all of the different varieties? (*2*)
(e) How can moulds be used to recycle waste? (*2*)
(f) Work out the energy chain for animal protein, starting with the sun and a sugar plant and ending with an animal eating the protein. (*2*)
(g) How will a sweet factory's plan for recycling waste improve the firm's budget? (*2*)
(h) When a fermenting tank is set up to grow mould on sugar waste, many conditions must be just right. Name two of these. (*2*)
(i) Selling processed mould for animal protein depends on at least two characteristics of the food. What might they be? (*2*)
(j) The mould used to process sugar waste is changing the kinds of food materials available into a more useful form. What is this change and what else would be needed by the mould to grow most efficiently? (*2*)

*42 A food-making machine: solving a space problem

What can spacemen eat? If they live on dried food all of the time, they will not have a balanced diet. They will need some fresh green food for vitamins, minerals and variety. This design is for a machine that will culture small algae called *Chlorella* in water.

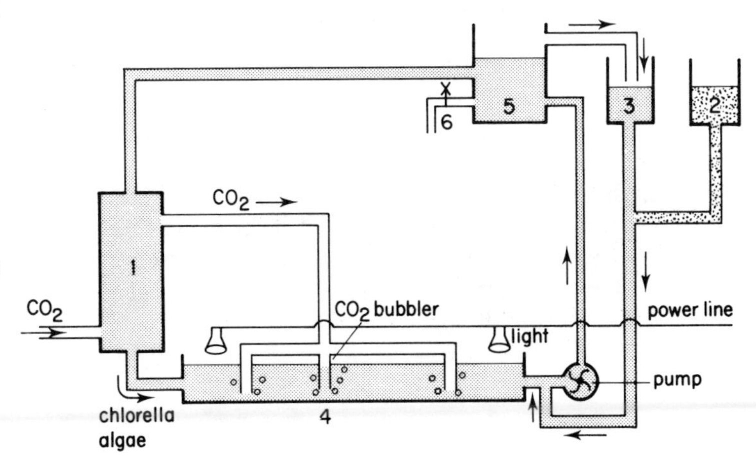

1 Gas exchanger tank, CO_2 in
2 Tank of mineral salts in water
3 Overflow and mixing tank
4 Culturing tank
5 Settling tank
6 Tap for drawing off algae

Chlorella plants are single-celled and each separate plant is thus very small. A tankful of them looks like green soup. *Chlorella* plants are green because they contain chlorophyll. They can carry out photosynthesis to join carbon dioxide and water to make sugar and starch. The chemical reactions of photosynthesis must begin in the light. Light energy is needed to split hydrogen from water molecules (formula H_2O); then the hydrogen can be joined to carbon dioxide molecules to give simple sugar. If the spacemen could strain and eat the *Chlorella* plants, they would get food from them. The plants would contain some sugar and some starch. If *Chlorella* plants are cultured for long enough, they have time to absorb minerals from the water and to build up some protein by combining the minerals with by-products of carbohydrate chemistry. So the spacemen would get some protein as well. The disadvantage of this method of food production is that *Chlorella* can only provide 10–12% of the human food ration. They also taste unpleasant.

(a) What is *Chlorella*? *(2)*
(b) What conditions are essential for a successful *Chlorella* culture? *(3)*
(c) Name the life process that *Chlorella* and all green plants use to capture light energy. *(1)*
(d) Which food material would these plants contain soon after the start of the culture? *(1)*
(e) How would added minerals improve the quality of food produced by *Chlorella*? *(1)*
(f) What must be done to the culture before spacemen would eat *Chlorella*? *(2)*
(g) How must the experimental set-up in the diagram be adapted for spaceflight? *(1)*
(h) A spacecraft is isolated in space, with only certain amounts of stored energy aboard. Which parts of this machine would use up some of the spacecraft's power? *(2)*
(i) Draw out an energy conversion chain to show how solar cells could provide energy for the spacemen in the form of green food from *Chlorella* culture. *(3)*
(j) If the *Chlorella* plants convert light energy with 10% efficiency, how much energy will they convert if they are given 1250 MJ of light at the best intensity for them? *(1)*

43 Egg production

A flock of laying hens was put into cages at the age of 20 weeks. They were kept for a year, until they were 72 weeks old. The number of eggs laid by the flock every week was recorded. These are the results in eggs per week against the age of the hens.

(a) Plot a graph of this information.

Age (weeks)	20	24	28	32	36	40	44	48	52	56	60	64	68	72
Eggs laid	0	18	68	88	88	83	78	76	73	70	68	63	58	53

(5)

(b) Write in your own words a description of the change in egg production with the age of the hens. *(3)*
(c) What was the age of the hens when they were giving the best level of production? *(1)*
(d) When do they show the greatest improvement in egg production? *(1)*

*44 Research on moulds

This is the story of the discovery of penicillin. Like so many great scientific discoveries, it was not made by one scientist, but by one scientist working on problems set by others. The other scientists had noted the effect of moulds on bacterial growth, but they did not follow up their ideas because they were working on another problem. The drug, penicillin, comes from a greenish-looking mould. It is quite common and its tiny spores are always floating in the air.

About 1877, Frenchmen Pasteur and Joubert observed that substances produced by moulds reduced the growth of other microbes. They pointed out the possibility of using moulds to treat infections in man and animals. No more work was done until 1928 when Alexander Fleming accidentally discovered that a mould, *Penicillium*, growing on one of his dishes of bacteria had prevented their growth.

In 1940, Howard Florey investigated *Penicillium* and grew enough mould to give an extract of penicillin. He tested the extract on mice. He infected mice with a germ called *Staphylococcus*. It gave the mice septic spots. He divided the mice into two groups, one with no treatment and one with regular treatment using penicillin. The untreated mice died. The mice which had regular treatment got better. Florey developed his work and penicillin was manufactured commercially and used in pills, ointments and injections. It saved many lives in World War II and Florey was given a knighthood for his work. He also won a Nobel Prize in 1945.

By 1959, penicillin seemed to become less effective. It was found that some bacteria had developed a resistance to the drug. Instead of penicillin destroying them, they were able to destroy the penicillin by digesting it. Work began on new strains of *Penicillium* and now over 5000 different kinds of penicillin are manufactured.

(a) Where does penicillin come from? *(2)*
(b) Who first noticed its effects? *(2)*
(c) Why did Florey use two groups of mice? *(2)*

(d) What happened after about twenty years of medical applications of penicillin? *(2)*
(e) How was this problem overcome? *(2)*
(f) How do moulds spread from one place to another? *(2)*
(g) A successful research plan will suggest new lines of research. Use examples from this paragraph to show how productive new ideas have arisen. *(1)*
(h) Name two technical problems that had to be overcome before penicillin could become a commercial success. *(2)*
(i) What kind of metabolic (physiological) change would have occurred to make bacteria penicillin resistant? *(2)*

*45 Oil fractions

Crude oil is a mixture of compounds containing carbon and hydrogen called hydrocarbons. The mixture can be separated into fractions of different boiling ranges by heating the mixture to different temperatures and collecting the fractions.

Fraction	*Boiling range (°C)*	*Number of carbon atoms in compounds*	*Amount in crude oil (%)*
gas	below 20	1–4	4
naphtha	40–180	4–12	20
kerosene	175–270	9–16	4
gas oil	200–400	15–25	26
fuel oil	above 300	20+	46

(a) Natural gas is mainly methane (CH_4). Which fraction is it found in? *(1)*
(b) Petrol is mainly octane (C_8H_{18}). In which boiling range is it found? *(1)*
(c) Diesel fuel is found in the gas oil fraction. What is the smallest number of carbon atoms in a diesel fuel hydrocarbon? What is the largest? *(2)*
(d) Plot a bar chart showing the percentage of each fraction (vertical axis) against boiling range (horizontal axis). *(5)*
(e) (i) Only half the fraction in which petrol is found is actually used for petrol. What is the percentage of crude oil used directly for petrol? *(1)*
(ii) In fact 30% of crude oil is used for petrol. The additional petrol is produced by breaking carbon compounds containing more carbon atoms than octane. Suggest the most suitable fraction to use. *(1)*
(f) Why should a fuel obtained from crude oil be burnt with plenty of air? *(3)*

46 Types of steel

Steel is mainly iron with small amounts of carbon added. The carbon alters some of the physical properties described in the table.

Strength	The maximum load that the steel can withstand without breaking
Hardness	Resistance of the surface to scratching and denting
Elongation	The maximum increase in length when the steel is stretched until it breaks

The diagram shows how these properties vary as the percentage of carbon in the steel is increased.

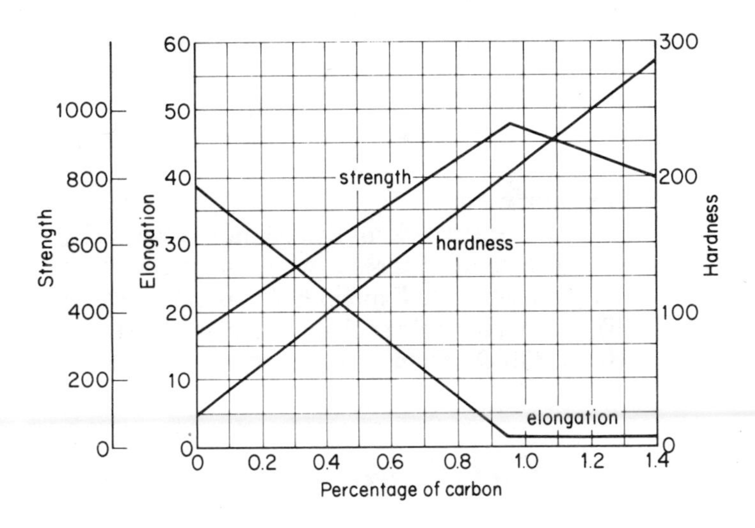

(a) What is the percentage of carbon which gives the steel with the highest strength? *(1)*
(b) For a steel containing 0.4% carbon, give its
(i) elongation,
(ii) hardness,
(iii) strength. *(3)*
(c) The steel used for car bodies contains 0.1% carbon. Suggest a reason why this steel is used. *(1)*
(d) Wire ropes are made from a steel with 0.6% carbon. Why is this steel suitable? *(2)*
(e) Suggest the percentage of carbon in the steel used to make hammers. *(1)*

47 Greenhouse

Mrs Green has a detached house. She had a greenhouse built against one side of it. She wanted to know how the temperature in the greenhouse changed during a sunny day because she wanted to grow tomatoes and it is important that the temperature should not be more than 28°C.

She set the maximum and minimum thermometer at breakfast time. At 8 o'clock the temperature was 21°C. An hour later it had reached 27°C. By 9.00a.m. the temperature was 30°C. An hour after that it was 34°C. At noon, the temperature was again 34°C. The temperature then fell by 2°C every hour until 3.00 p.m. At 4.00 p.m. the temperature was 25°C and two hours later it had fallen to 20°C. The highest reading on the 'maximum and minimum' thermometer was 36°C. The sun shone all day.

(a) Put this information in a table. *(5)*
(b) Draw a graph to show how the temperature changed during the day. *(5)*
(c) Estimate the time at which the highest temperature was reached. *(1)*
(d) On which side of the house was the greenhouse built? *(1)*
(e) Mrs Green decides to have a fan fitted to suck hot air out of the greenhouse when the temperature is more than 25°C. It would cost 1p per hour to operate the fan.
   (i) Mrs Green reckons that on the day when she took temperatures it would have cost 7.33p to use a fan. Explain her reasoning. *(2)*
   (ii) In fact it would probably have cost less than 7.33p. Explain why. *(2)*

48 Pest control

Screw-worm flies used to be a problem in the USA, but in 1959 they were exterminated by agricultural officers using gamma radiation.

The female flies lay eggs in any scratches or wounds on cattle. The resulting maggots feed on the wound and make it bigger. Cattle often die as a result.

The female mates only once before laying eggs. If the male is sterile, the eggs will not hatch.

Male screw-worm flies can be sterilized by gamma radiation. These can be released in the area where screw-worm flies are a nuisance. If there are as many sterile males as wild males, half the batches of eggs will not hatch. So the next generation will be only half the size that it would have been if no sterile males had been put into the area.

Male pupae are sterilized by 25 gray of gamma radiation. Females need 50 gray. These doses do not kill them.

Six 4.85×10^{13} bequerel ^{60}Co γ-ray sources were set up in an aeroplane hangar in Florida. They were used to give each pupa 80 gray. 50 million sterile flies were produced each week. These were dropped by 20 planes over 180 000 km^2 of the South Western USA. After six months it was impossible to find any living maggots in the area.

(a) More radiation is needed to sterilize female pupae than male ones. How much more? *(3)*
(b) What was the total strength of the radioactive sources in the aircraft hangar? *(3)*
(c) About how many sterilized flies were produced during the project? *(3)*
(d) How many flies were loaded on to each plane? *(3)*
(e) (i) How many flies on average were put into each square kilometre of the area each week? *(3)*
   (ii) A farmer, Mr MacDonald, wanted to know how many sterile flies were on his 1 km^2 farm. The agricultural officers were unable to tell him. Why is this? *(1)*
(f) Why are pupae given 80 gray of radiation when 25 gray is enough to sterilize males? *(2)*
(g) A female fly may produce on average 20 eggs. Even if these are not sterile it is likely that only two will hatch and survive to breed successfully.
   (i) If there are 1000 flies in an area (500 male, 500 female), how many eggs are likely to be produced as the next generation? *(2)*
   (ii) How many of these eggs are likely to survive to breed? *(2)*
   (iii) If the original 1000 flies are joined by 1000 sterile flies (500 male, 500 female), how many eggs are likely to be produced as the next generation? *(2)*
   (iv) How many of these eggs, produced by these 2000 flies are likely to hatch and survive to breed? *(1)*
   (v) Explain what happens if the surviving flies from (iv) are joined by another 1000 sterile flies. *(3)*

2 Problem solving

A

1

Ants like dry places.
Ants like warm places.
Ants nest in the ground or in old logs and stumps.
Ants do not like paraffin or cinders.
Ants eat sugar, cake and dead bodies (and a lot of other organic matter).

(a) Mrs McGrath comes home to find hundreds of ants marching into her kitchen. The cause of this problem may be
A. it is much warmer outside than in the house.
B. sugar has been spilt on the floor.
C. there is a cream cake in the freezer.
D. paraffin is stored in the kitchen.

(b) She needs to stop more ants coming into the kitchen. One way to do this would be to
A. take the paraffin which is stored in the kitchen out to the garage.
B. put sugar across the doorway.
C. put cinders across the doorway.
D. put the heating on.

*2

The diagram shows Mrs Timms' shop. She stands behind the shop counter.

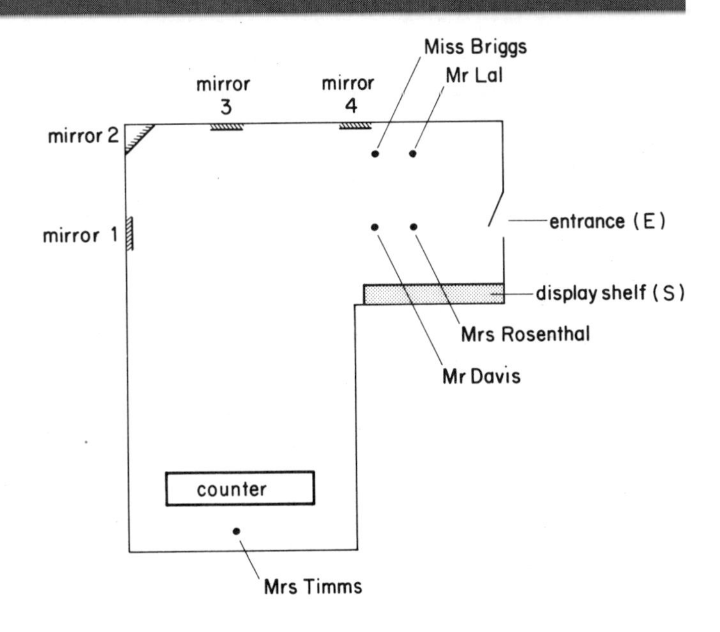

(a) There are four customers in the shop. Which customer cannot be seen by Mrs Timms without looking in a mirror?
A. Miss Briggs.
B. Mr Lal.
C. Mr Davis.
D. Mrs Rosenthal.

4.3d

(b) Which of the mirrors could be used by Mrs Timms to see the entrance of the shop, E?
A. Mirror 1.
B. Mirror 2.
C. Mirror 3.
D. Mirror 4.

(c) Which of the mirrors could be used by Mrs Timms to see customers at the display shelf, S?
A. Mirror 1.
B. Mirror 2.
C. Mirror 3.
D. Mirror 4.

3

Elna Smölander had an 'organic' market garden. She wanted to stop the caterpillars of cabbage white butterflies eating the cabbages without using pesticides. She counted the number of caterpillars on a square metre of ground planted with various crops:

	Cabbages	*Cauliflowers*	*Tomatoes*	*Strawberries*	*Potatoes*
Caterpillars	30	36	0	5	15

Spraying cabbages with one of the following liquids worked well. Which do you think it was?
A. Water in which cauliflower has been boiled.
B. Water in which tomatoes have been squeezed.
C. Water containing strawberry juice.
D. Water in which potatoes have been boiled.

4

The wire for an electric heating element ought to have the following properties:

(i) it should resist corrosion at high temperatures,
(ii) it should have a high melting point,
(iii) it should have a moderate electrical resistance.

Which of the following wires would be the most suitable for the heating element of an electric fire?

	Electrical resistance	Resistance to corrosion	Melting point
A.	high	good	high
B.	moderate	moderate	high
C.	moderate	good	moderate
D.	moderate	good	high

*5

A factory makes door knobs, ashtrays and other products from a metal alloy using a casting machine. It is thought that there was some grit in the metal put into the machine on one day. Grit has a density of about 2.5 g/cm^3. The table shows the average mass and volume of products made during part of the week.

Day	Volume (cm^3)	Mass (g)
Monday	200	1260
Tuesday	300	1770
Wednesday	500	3150
Thursday	700	4410

(a) What is the density of the metal?
  A. 0.16 g/cm^3.
  B. 5.9 g/cm^3.
  C. 6.3 g/cm^3.
  D. 15.8 g/cm^3.

(b) Which day did grit get into the metal?
  A. Monday.
  B. Tuesday.
  C. Wednesday.
  D. Thursday.

*6

A

Local environmental health regulations require that waste water put into public drains should have a pH of 8–9. A spot check on a company's waste water showed its pH was 3. Which of the following would be the best to correct the pH of the waste water?

A. ammonia solution.
B. calcium carbonate (solid).
C. sodium hydroxide solution.
D. magnesium hydroxide (solid).

*7

The waste gases from coal and oil fired power stations contain sulphur dioxide.

(a) Sulphur dioxide must be removed because it can cause
  A. radioactive contamination.
  B. lead poisoning.
  C. smog.
  D. acid rain.

(b) The amount of sulphur dioxide in the waste gases could be reduced most effectively by passing the waste gases through
  A. calcium hydroxide solution.
  B. sodium hydroxide solution.
  C. solid sodium hydroxide.
  D. water.

8

A

Some pupils wanted to test various types of dye to see which was best at dyeing a fabric. They dipped pieces of the fabric into the different dyes, washed and dried them. The pieces of fabric were then compared.

(a) What three things should really be kept the same if this is to be a fair experiment? *(3)*
(b) What could you actually measure to decide the best? *(1)*
(c) What should be done to make sure the results are reliable? *(1)*

*9

Some pupils wanted to test various types of bath crystals to see which was best at making water soft. Soft water produces more bubbles (lather) when soap is added to it. They added some bath crystals to cold water. Some washing-up liquid was then added and the water shaken.

(a) What three things should really be kept the same if this is to be a fair experiment? *(3)*
(b) What could you actually measure to decide the best? *(1)*

10

A pupil knew that some plants spread their fruit by sticking on to passing animals. 'Goosegrass' is an example. It has small round fruit covered in tiny hooks. This pupil wanted to test whether they clung better on to fur or wool. He decided to put a few of these on to some fur and a few on to some wool. He then shook these materials for a while.

goosegrass

(a) Write down three things he should keep the same if he is to do a fair test. *(3)*
(b) Describe how you would measure which material the fruit sticks to best. (2)

*11

Some pupils wanted to compare the effect of dilute acid on the bones of different animals. They put acid into a number of beakers. They put one bone from an animal into each beaker. They were left in the laboratory for some time. The pupils came back and checked to find out if any had been softened by the acid.

(a) If this is to be a fair experiment, what three things should they have kept the same? *(3)*
(b) What could they measure to find out which animal's bones were most affected by the acid? *(1)*

12

Some pupils had a number of solid materials which would burn. They wanted to know which was the best fuel. They were going to measure the amount of ash that each fuel left behind when burnt. The one with the least ash would be the best.

(a) How could they measure the 'amount of ash' left and what instrument would they need to measure it? (2)
(b) When doing this experiment it has to be a fair test. What two things should be kept the same? (2)
(c) There are several ways of deciding on what is a perfect fuel. Name any two of them (other than catching alight easily and the amount of ash left). (2)

13

A pupil has several bricks. She wants to find out which one soaks up the most water if wet. This is her plan.

(a) Take pieces of brick of equal size.
(b) Measure equal volumes of water into buckets.
(c) Put one piece of brick into each bucket.

What measurements and calculations should she then make to find out which holds most water?

14

A pupil wants to find out which of three methods is best for preserving broad beans. He uses the same amount of beans and preserves them in the three ways. They are left in the same conditions for one year. What measurements and calculations should he then make to find out which method is the best?

15

Most people prefer poached eggs which have a firm white. The formation of a firm white may depend on the pH of the water used. The pH of the cooking water can be adjusted by adding substances found in the kitchen. Design an experiment to find the best conditions to get a firm white part of a poached egg. Materials and equipment available: pH meter; bunsen burners, tripod and gauzes; beakers, 250 cm^3 and 400 cm^3; test tubes; eggs; sodium chloride (salt); sodium hydrogen carbonate (bicarbonate of soda); sucrose (sugar); ethanoic acid (vinegar); dropping pipettes.
Note: You need not use all the equipment.

16 Baking powders

In making cakes, baking powder is added to flour. The active ingredients of baking powder are sodium hydrogen carbonate and tartaric acid. When water is added, the active ingredients react and carbon dioxide is produced. The carbon dioxide helps the cake to rise when it is cooked.

You are asked to carry out a best buy survey of the baking powders which can be bought in the local supermarkets. Suggest the experiment(s) you would perform to help you decide which is the best buy. Your answer should include the details of how you would do the experiment(s). Assume usual laboratory equipment is available and also a top pan balance.

*17

Pollen from the anther of a flower on one plant can get to the stigma of a flower on another plant. This is called cross pollination. New varieties of plants can then be produced as a result. In nature, pollen is carried by insects, birds or simply by the air. A rose grower wants to produce new varieties of rose. He selects two particular plants A and B for cross pollination.

(a) How could he get pollen from plant A to plant B? *(1)*

(b) How could he make sure that no pollen from the anther of the flower on plant B reached the stigma of the same flower? *(1)*

(c) How could he make sure that no pollen from any other flowers reached the stigma of the flower on plant B? *(1)*

*18

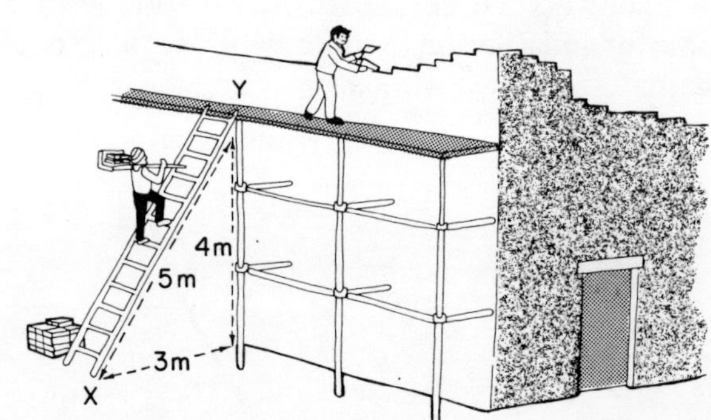

A man has to move some bricks from X to Y. They have a weight of 200 N. How much work is needed? Use the formula: work = force × distance moved.

19

Some lenses were in marked envelopes but they have been mixed up by mistake. You have to put them back into the correct envelopes. The picture shows what a letter (I) looks like when using the lenses.

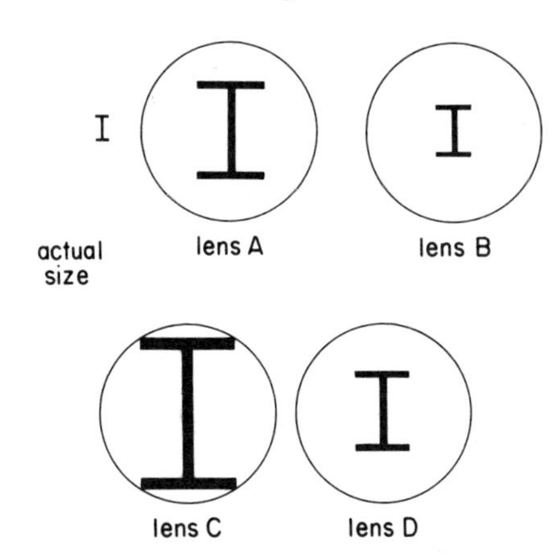

(a) Put them in order with the biggest magnification first. *(2)*
(b) Using your ruler, how many times does lens A magnify the letter 'I'? *(1)*
(c) How many times does lens C magnify? *(1)*
(d) What two things about the 'I' are made to look bigger when you use the lenses? *(1)*

20 Catching thieves

This question is about evidence and how it can be used. You are a forensic scientist and you have discovered these clues. Some are from the scene of the crime and some are from a suspect. How would you use this evidence?

Clue 1
There was a footprint in the flower bed under the window of a burgled house. How could this be used as evidence? *(2)*

Clue 2
In the hem of the suspect's trousers, there was some dust. When you put the dust under the microscope, you found some green fibres. How could you use these in evidence? *(2)*

Clue 3
Mud on the suspect's trousers showed a high proportion of chalk. Could this be useful evidence? *(2)*

Clue 4
Metal filings were in the dust from his trousers. A marked gun was one of the items stolen from the house. Could these filings tell you anything? *(2)*

Clue 5
There is one clue that could be a final proof, but you would have to look for it at the scene of the crime. Say what it could be and how you would collect it. *(2)*

21

Rashmi wants to find the energy value of betel nuts.

Energy value = Energy (joules) produced by completely burning the nut ÷ Mass of nut (kilograms)

She burns the nut under a dish of water. The dish is so thin that it absorbs very little energy from the burning nut. The energy given to the water can be calculated using the formula:

Energy = Mass of water (in kg) × Specific heat capacity of water (in J/kg°C) × Temperature rise (in °C)

The specific heat capacity of water is 4200 J/kg°C. Rashmi records these results:

Mass of nut	= 0.005 kg
Mass of dish full of water	= 0.150 kg
Final temperature of water	= 35°C

(a) She needs two more pieces of information. What are they? *(2)*
(b) Rashmi may have a problem with this experiment. Betel nuts are quite big. The outside of the nut burns. Then it might stop burning. How can she cope with this? *(2)*

22

Some people say, 'Paper containers are better than plastic ones.' There are several things that they could mean by this statement. Some of these could never be checked out scientifically.

(a) Name three things that really could be checked. *(3)*
(b) Explain briefly how each of these things could be tested. *(9)*

23

You are watching a TV commercial and it says, 'Cleano is the best toilet cleaner'. There are several things that they could mean by this statement. Some of these could never be checked out scientifically.

(a) Name three things that really could be checked. *(3)*
(b) Explain briefly how each of these things could be tested. *(9)*

24

Some people say, 'Clothes made from synthetic fibres are much better than those made from natural fibres.' There are several things that they could mean by this statement. Some of these could never be checked out scientifically.

(a) Name three things that really could be checked. *(3)*
(b) Explain briefly how each of these things could be tested. *(9)*

25

You are asked to test two waterproofing sprays for tent material. Design a fair test to decide which spray is the most effective.

(a) Draw the test in action. *(4)*
(b) Give clear instructions about how to carry out the test. *(4)*
(c) Explain how you would evaluate your results. *(2)*

26

Design an experiment to compare the elasticity of five threads. The elasticity of a thread is how much it stretches when pulled by a particular force.

(a) Draw a labelled diagram to show the apparatus you would use. *(3)*
(b) Explain how you would do the experiment. *(3)*
(c) Describe the kind of results which you would expect. *(2)*
(d) (i) When would you choose to use an elastic thread or fabric? *(1)*
(ii) When would you choose to use an inelastic thread? *(1)*

27 The flammability of fabrics

All fabrics will eventually catch fire and burn. Below are listed different properties associated with the burning process.

(a) How easily it catches fire.
(b) The length of time it takes for the fabric to burn.
(c) The amount of heat given out while it burns.
(d) The way the fabric burns, for example melts and then catches fire.

For each property, devise a suitable test so that various fabrics can be compared.

*28 Parking light

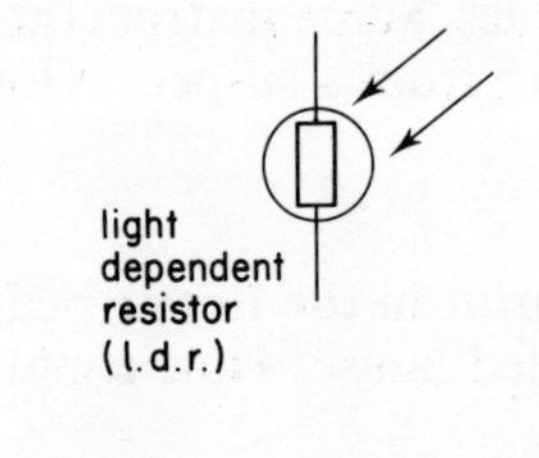

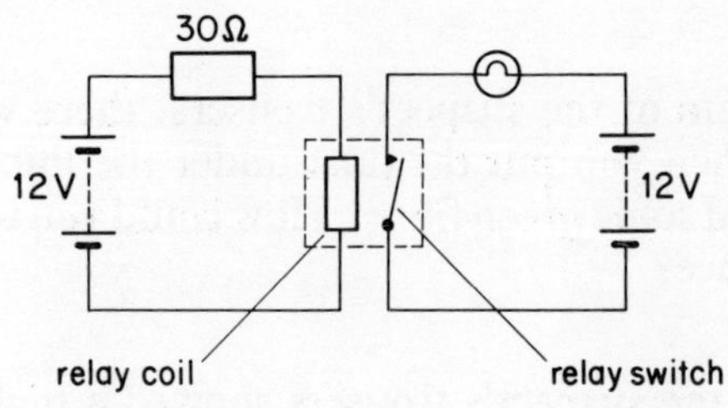

Mr Lee wants to make the parking lights on his car come on automatically when it goes dark. In the circuit shown, the potential difference (voltage) across the relay coil is 6 V. This potential difference is just enough for the electromagnetic force from the relay coil to close the relay switch. This allows the filament lamp to come on.

The light dependent resistor (l.d.r.) has a resistance of 300 Ω in the dark and 30 Ω in bright light.

(a) Explain how the circuit works if the 30 Ω resistor is replaced by the l.d.r. (3)
(b) Redraw the circuit showing where Mr Lee should connect the l.d.r. so that the filament light comes on when it goes dark. (2)

29 Woodwork

Diagram 1 shows a block of wood. When it is dry

$x = 10.0$ cm; $y = 10.0$ cm; $z = 10.0$ cm

When it is wet

$x = 10.5$ cm; $y = 10.0$ cm; $Z = 10.5$ cm.

Diagram 1

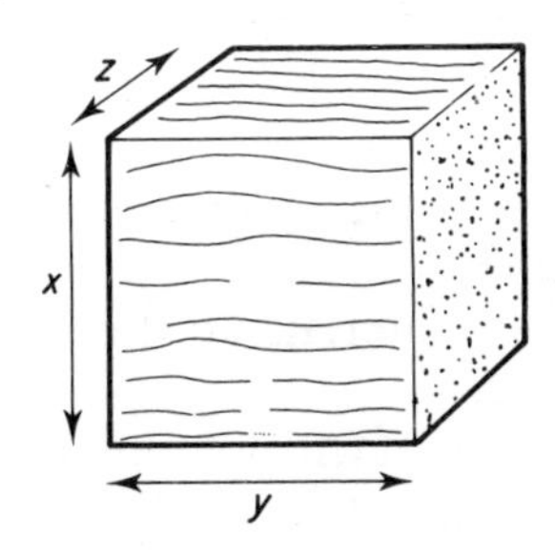

The ancient Egyptians had a problem cutting the stone for Cleopatra's Needle (a block of stone about 10 m long, 1 m thick and 1 m wide). It is said that they cut grooves in the stone and then used wood to split the stone. Explain exactly what they should have done. Which way should the grain of the wood be arranged? Would thin pieces of wood be better than thick ones?

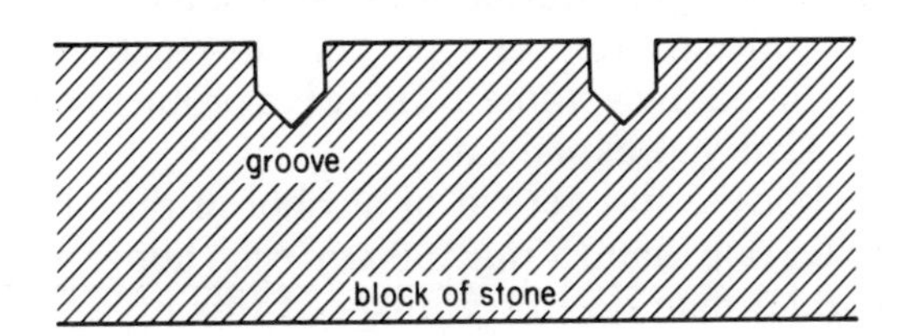

30 Bee vision

Scientists noticed that if a real and an artificial flower are put together in sunlight, bees go to the real one only. If this is done indoors under an ordinary electric lamp, bees do not go to either of the flowers. The scientists took some black and white photographs to try to solve the problem. They used a special film.

Diagram 1

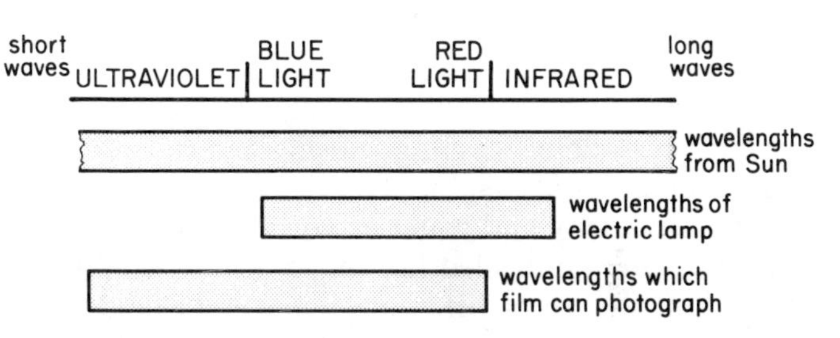

Diagram 1 tells us about the wavelengths in sunlight and electric light. It also shows which wavelengths could be photographed by the film. Diagram 2(a) shows a photograph of a coreopsis flower taken in electric light. Diagram 2(b) shows a photograph of the same flower taken in sunlight. Use this information to work out how bees are attracted to flowers. Explain all your reasoning.

Diagram 2

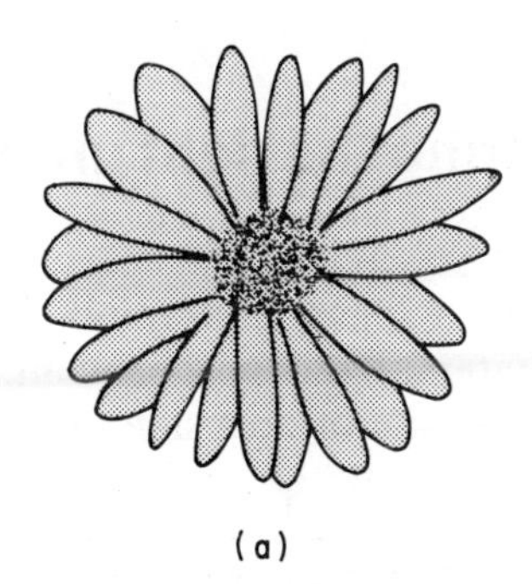
(a)

(b)

A

31 Domestic appliances

Cookers are already equipped with simple electrical controls for time and temperature, but an electronic system could be programmed to select for itself a suitable cooking plan when given the recipe and the time the dish is to be eaten. It may be that our simple cookers will be replaced by cooking machines equipped with automatic hoppers to supply common ingredients in measured amounts direct into the cooking pots. The freezer and fridge would be part of the design. The machine would be able to thaw and cook by microwave if necessary.

Washing machines could be programmed to select the correct cycle of washing, rinsing and drying for different mixes and weights of fabric. They would use as little electricity and water as possible.

A computer-controlled knitting machine would be able to knit all sorts of different stitches and would be able to adapt a pattern to the measurements of the wearer. It would be able to use several different kinds of wool at once and it could knit according to a punch card or a preselected programme.

Hairdressing could be computer controlled. Waving machines and hair colouring machines would be preset to the condition of the customer's hair and the chosen perm or colour.

Choose one of these machines and design its control panel. Take care to include all the controls.

*32 Getting the right uranium

An American factory separates uranium-235 for use in nuclear power stations from the more common isotope uranium-238. The rate of diffusion of a molecule is proportional to

$$\frac{1}{\sqrt{(\text{relative molar mass})}}$$

This is Graham's Law.

Uranium hexafluoride is a gas. The separation process depends on the fact that molecules of uranium hexafluoride (UF_6) containing U-235 diffuse through a barrier faster than those containing U-238.

The factory needs 4000 separate diffusion barriers and covers two and a half square kilometres. What is the problem? (You may wish to make comparisons with a similar process for separating oxygen (O_2) and nitrogen (N_2).

Atom	*Relative atomic mass*
uranium-235	235
uranium-238	238
fluorine	19
oxygen	16
nitrogen	14

A

*33 A fair cop?

In Britain, the legal limit of alcohol in a driver's blood is 80 mg per 100 cm^3. It is thought that the body gets rid of alcohol in the blood at the rate of 15 mg per 100 cm^3 each hour. The police have sometimes used this information to work out alcohol levels an hour or two before blood samples were taken.

A man is stopped by police on the M6 motorway near Birmingham. The speed limit on the M6 is 70 m.p.h. He says that he has driven from Lancaster. Lancaster is 140 miles north of Birmingham. He cannot remember what time he left home. He has alcohol in his blood but the level is only 55 mg/100 cm^3.

(a) He may have a problem. Explain this. *(6)*

(b) To start with, he tells the police that he had a few pints of beer in Lancaster. In court, he 'remembers' that he came off the motorway about an hour before Birmingham. He had another drink in a public house. How does this help him? *(2)*

(c) (i) What is the volume of blood in an average adult? (1)

(ii) What materials do the kidneys remove from blood? *(1)*

(iii) 125 cm^3 of whisky contains 50 cm^3 of alcohol. 1000 cm^3 (1 litre) of beer contains 50 cm^3 of alcohol. If a driver does not want to drink more than 50 cm^3 of alcohol, should he be advised to drink beer or whisky? Is there any difference? Explain your answer. *(3)*

(d) Some people are not 'over the limit' after drinking 1 litre of beer.

(i) Estimate the concentration of alcohol in the blood (by volume) if all the alcohol is taken into the blood stream. *(1)*

(ii) The legal limit of 80 mg/100 cm^3 is a concentration of 1 part alcohol to 1000 parts blood (by volume). There is a problem here. Can you explain it? *(2)*

(e) The body could get rid of alcohol in the blood at a faster rate than 15 mg/100 cm^3 each hour. What should a man do to try to speed this up? *(2)*

*34 Market garden

This question is about the problems of running a market garden. Study the flow diagram.

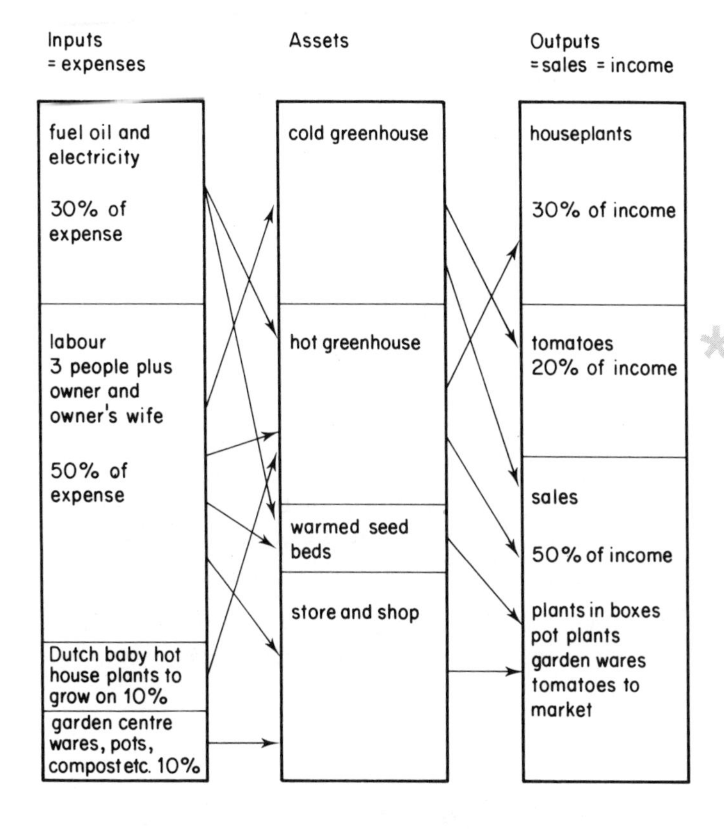

(a) Some market gardens have to close down when there is an increase in the price of oil. Why is this when electricity is more expensive? *(2)*

(b) Which of the expense inputs could be kept low by bulk buying? *(2)*

(c) Would it be wise to recycle soil from the germination beds to use in cold greenhouses? Give reasons for your answer. *(2)*

(d) Chrysanthemums will only flower under glass in special lighting conditions. This means that for a Christmas crop, plants are brought on by extra lighting controls. The fuel bill for this is equal to 1/6th of an employee's wages. How would you make sure of a profit on chrysanthemums? *(2)*

(e) Tomato seedlings are started in warmth in February. The warm greenhouse is used to produce pot plants for Christmas but some of these will still be left after Christmas. How would you manage the warm greenhouse to cope with pot plants and tomato seedlings during the winter? *(2)*

(f) Give three good reasons for employing one person to run the garden centre full-time. What could they do in their spare time? *(2)*

(g) The business could be improved by having a greater variety of things for sale. Of course these would need to be things which people were interested in buying. How would you do this without too much increase in expenses? *(2)*

(h) Electronic systems could aid this business. Describe two such useful systems and how they would be effective. *(2)*

(i) Would cutting down on baby hothouse plants from Holland have a good effect on the business? *(2)*

*35 Train journeys

Mombasa is about 500 km from Nairobi. Mombasa is on the coast. Nairobi is at a height of 1500 m. The average speed of a train travelling from Mombasa to Nairobi is 40 km/h. The average speed of a train travelling from Nairobi to Mombasa is 50 km/h. The two towns are joined by a single track railway line. Two passenger trains operate on the line. A train leaves each town at 8.00 a.m. daily. The distance/time graph shows the journey from Nairobi to Mombasa.

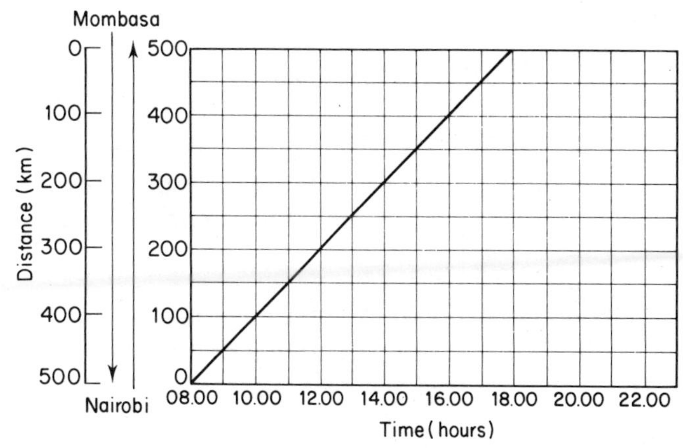

(a) (i) Copy it. *(1)*

(ii) Draw another line on the graph to show the journey from Mombasa to Nairobi. *(3)*

(iii) Find out where a passing place must be provided on the single-track railway. *(2)*

(b) Explain in terms of potential energy and work done against friction why the train has a greater problem in getting from Mombasa to Nairobi than in going the other way. *(5)*

*36 Kitchen costing

The data in the table is for electrical equipment used in a family kitchen in an average week.

Appliance	*Power rating* (kW)	*Weekly use* (hour)
electric cooker	7.0	5
microwave cooker	1.3	4
slow cooker	0.2	20
dishwasher	2.5	10
electric kettle	2.0	2
refrigerator	0.5	84 (with thermostat)
deep freeze	1.0	84 (with thermostat)
washing machine	2.5	12

(a) Given that power (in watts) = current × voltage and that 1 A, 3 A, 5 A, 10 A and 13 A fuses are available, work out the most suitable fuse for the plug on each appliance. (Mains voltage is 240 V) *(8)*
(b) What fuse wire should be used for the electric cooker circuit? *(1)*
(c) A unit of electricity is a kilowatt hour (kWh). If 1 kWh costs 6p, how much does it cost to run the kitchen for a week? *(4)*

*37 Turbine trouble

The tensile strength of a material depends on its temperature as shown by the graph. (Tensile strength is the biggest force that can be put on a square metre of a material without it breaking.)

Engineers have a rule: You cannot use a metal machine part at a temperature higher than half its melting point in degrees Kelvin.

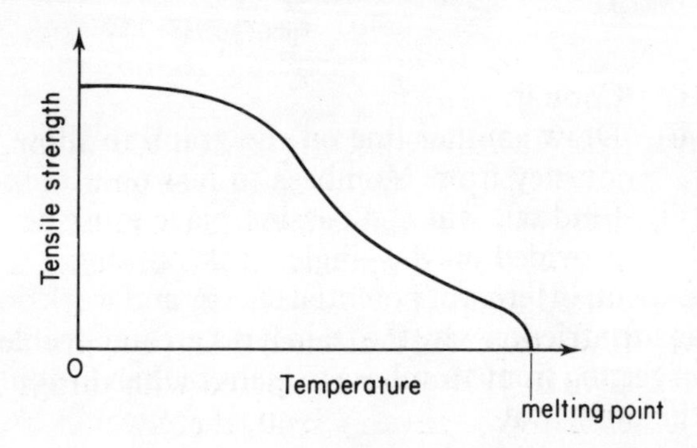

$$\text{Temperature in degrees Kelvin} = \text{Temperature in degrees Celsius} + 273$$

The highest working temperatures are about 950 °C for turbine blade alloys in aircraft.

(a) What is the melting point of these alloys in °C? *(3)*
(b) (i) Copy the graph. Assume that it is for the turbine blade alloy. Mark 950°C at an appropriate point on the temperature axis. *(2)*
(ii) Explain what problems would occur if the aeroplane engine overheated to a temperature of 1150°C. *(2)*

*38 Electroplating

The Fabjewel Company makes expensive jewellery. One style is made of copper which is then coated with nickel. The copper is worked into the required shape and heated to remove the stresses that occur during the shaping process. There is a film of copper(II) oxide on the surface of the copper and also some grease.

(a) Devise a system which could be automated to:
(i) remove the grease and copper(II) oxide,
(ii) electroplate the nickel on to the copper surface,
(iii) dry the plated nickel to give a surface which is clean and free from nickel oxide. *(12)*
(b) Explain how pollution of the atmosphere and drains can be reduced or prevented. *(10)*
(c) Present this system as a flow chart. *(5)*

*39 Making the most of ammonia

Ammonia is produced in a chemical works by the Haber process. The process combines nitrogen and hydrogen at high temperature and pressure. The works manager wants to produce as much ammonia as possible each week. The graphs show how the speed of reaction and yield depend on temperature.

The amount of ammonia produced each week is proportional to the rate of reaction. It is also proportional to the yield. So it is proportional to Q where

Q = (rate of reaction) × (yield)

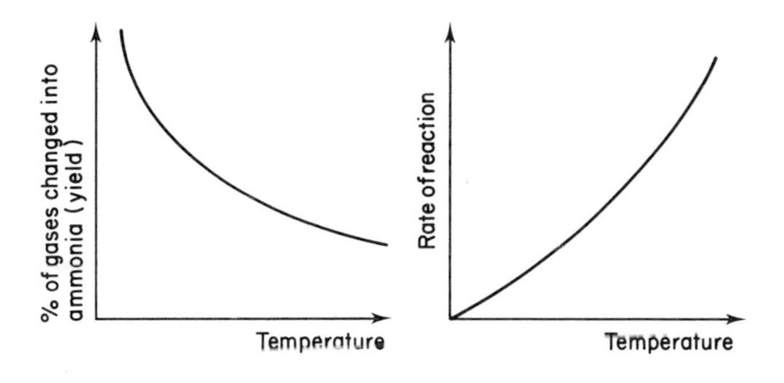

(a) Sketch a graph to show how Q depends on temperature. Show on the temperature axis the temperature which the works manager should use. *(4)*

(b) The temperature selected is usually about 500°C. What are the problems of keeping a reaction tank at this temperature? How can they be solved? *(6)*

*40 Carbon dating

The half-life of the isotope C-14 is 5700 years. This fact is used by archeologists to estimate the age of bones, trees, etc. which have been dug up. The bones of a man who died 5700 years ago should contain only half as much C-14 as the bones of a man who died recently. At least, that is the simple theory. Unfortunately it is a bit more complicated.

The graph shows how the percentage of C-14 remaining decreases with the age of the material.

Diagram 1

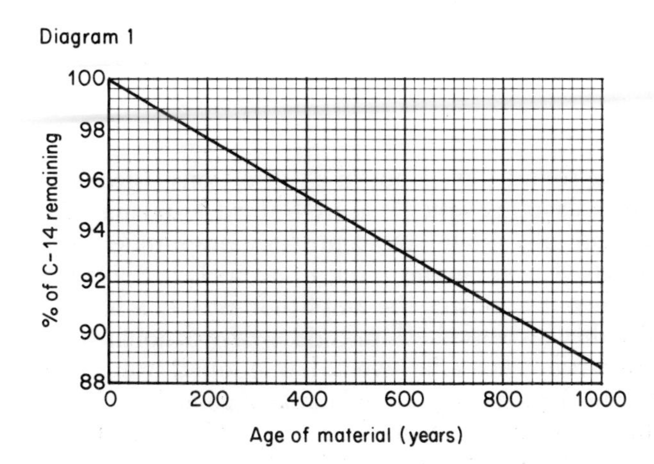

The level of C-14 in the atmosphere has been constant for millions of years. But coal and oil have been burned in factories and power stations during the last hundred years. This has produced a great deal of non-radioactive carbon. The concentration of C-14 in the atmosphere during the first half of this century was about 2% lower than before. However, the testing of nuclear weapons in the atmosphere between 1945 and the Test Ban Treaty in 1962 is said to have produced 3½ tonnes of C-14.

Living things, including trees and human beings, will gradually take in more and more of this C-14. The graph below is based on estimates in 1962. It shows how the amount of C-14 in living things was expected to vary in later years.

Diagram 2

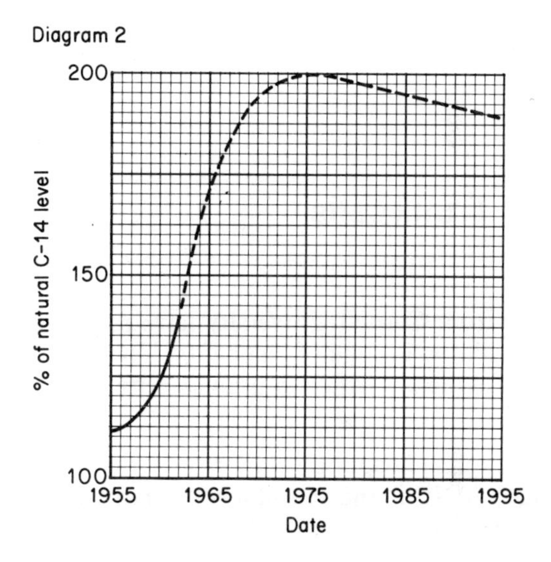

(a) (i) By what percentage has the amount of C-14 fallen in something which has been dead 600 years? *(2)*

(ii) C-14 is being made in the atmosphere all the time because of cosmic rays coming from outer space. So why has the amount of C-14 in the atmosphere not been rising throughout the last few million years? *(1)*

(iii) What has happened to the total amount of carbon in the atmosphere during the last hundred years? *(1)*

(iv) In 1850, the average tree took in 100 units of C-14. In 1950, the average tree took in 98 units of C-14. Explain this change. *(1)*

(v) Why is part of the second graph shown as a dotted line? *(1)*

(vi) It was expected that when things which were alive in 1970 were compared with things which were alive in 1960, there would be a rise in the amount of C-14 in them. Explain this. (2)

(vii) It was expected that when things which are alive in 1995 are compared with things which were alive in 1985, there would be a fall in the amount of C-14 in them. Explain this. (2)

(b) Joe Watson is an unscrupulous physicist who has decided to run an antiques business. He thinks it might be possible to use this information to make fake antique furniture which could pass the C-14 dating test. This would make a huge profit.

Describe how Joe might go about it. Explain precisely how you use each of the pieces of data supplied. Explain what period Joe should try to fake (that is, 1500–1600, 1700–1800, etc.) Explain carefully how he should choose the wood he uses. (8)

*41 Trunk calls

Scientists from Cornell University have recorded sounds made by groups of elephants. These have a very low frequency which is usually between 14 Hz and 30 Hz.

(a) The graph in Diagram 1 shows the human hearing range; 60 dB is the sound level of normal human conversation.

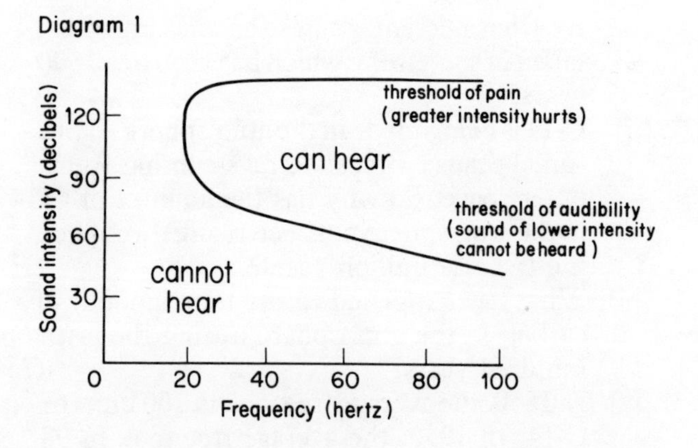

The recording of the sound made by the elephants is on a disc which can be put on a record player. The record player has a 'volume control' knob which can alter the sound level. It has three speeds: 33 r.p.m. (revolutions per minute), 45 r.p.m. and 78 r.p.m. The speed can be changed using the speed control knob.

To reproduce the sound made by the elephants the record is played at 33 r.p.m. The sound level is 60 dB but nothing can be heard.

(i) Explain how the record player should be adjusted in order to hear sounds of 30 Hz. (3)

(ii) Explain how the record player should be adjusted to change 14 Hz elephant talk into sounds which human beings can hear. (3)

(b) The velocity of sound is about 330 m/s. Use the formula

$$\text{Velocity} = (\text{frequency}) \times (\text{wavelength})$$

to calculate the greatest and smallest wavelengths produced by the elephants. (5)

(c) Scientists have a problem: elephants can hear best at 1000 Hz, so why do they talk at 14 to 30 Hz? Study the data below, then try to solve the problem.

Sound intensity is lower when sound is coming from further away.

1 Assuming air itself does not absorb sound

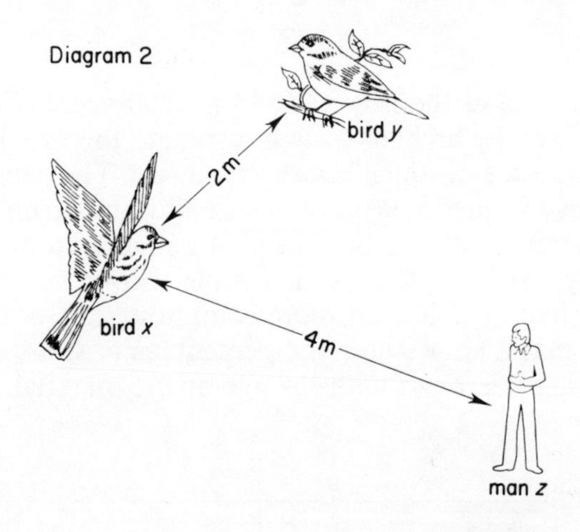

The sound travels out in all directions. Bird X produces sound which is heard by bird Y and man Z. We can talk about 'sound pressure amplitude' in pascals (Pa) or 'sound level' in decibels (dB). If the distance is doubled, the amplitude is only a half of what it was before. Amplitude is proportional to 1/distance. This is the inverse law of sound attenuation. So if bird Y gets 20×10^{-4} Pa, man Z gets 10×10^{-4} Pa. But what really matters is the sound level in dB. The sound level for bird Y is 40 dB and for man Z it is 34 dB. The sound level goes down by 20 dB if the distance becomes 10 times bigger. The table shows how these things change with distance.

Distance from bird (m)	2	4	6	8	10
Sound pressure amplitude ($\times 10^{-4}$ Pa)	20	10	6.7	5	4
Sound level (dB)	40	34	30.5	28	26

This effect does not depend on the frequency of the sound. All frequencies are affected in the same way.

2 Absorption of sound energy by air

This is mostly due to air friction (viscosity). Conduction of heat between high pressure areas and low pressure areas caused by sound is also important. But the effect of air friction is three times greater than the effect of conduction.

As long ago as 1845, Sir George Stokes proved that because of air friction the sound pressure amplitude is halved when distance increases by x where

$$x = 7 \times 10^3 \text{ (wavelength)}^2$$

The sound level will go down by 20 dB if the distance is increased by y where

$$y = 23 \times 10^3 \text{ (wavelength)}^2$$

The distances in these formulae assume that sound is travelling in one direction. The effect of spreading out (inverse law) is ignored. The formulae tell us about how sound penetration depends on wavelength.

(a) Assume that air itself does not absorb sound. Draw graphs for sound which has a pressure amplitude of 20×10^{-4} Pa and sound level of 40 dB at 2 m from the source.
(i) Show how pressure amplitude changes as distance increases from 2 m to 16 m. *(3)*
(ii) Show how sound level changes as distance increases from 2 m to 200 m. *(3)*

(b) *Absorption of sound energy by air, ignoring the inverse law.* Draw graphs for sound which has a sound level of 40 dB to show how sound level changes with distance between 0 m and 69×10^3 m (69 km). Use the same set of axes for both graphs:
(i) when the wavelength is 1 m, *(3)*
(ii) when the wavelength is 0.5 m. *(2)*

(c) Elephants often go round in groups but sometimes one gets separated from the others by several kilometres. Explain why you think elephants might prefer to talk at 14 Hz to 30 Hz rather than 1000 Hz. *(5)*

3 Patterns

1

Sparrows have short, thick bills for cracking seeds. Terns have dagger-shaped bills for catching fish.

Diagram 1

Diagram 2

Here are pictures of four birds.

(a) Which of these could be a kingfisher? (*1*)
(b) Which of these could be a corn bunting? (*1*)

A 2

Ammonia (NH_3) is produced industrially by reacting nitrogen and hydrogen.

$$N_2 + 3H_2 \rightleftharpoons 2NH_3$$

The reaction does not go to completion, that is, there is always some nitrogen and hydrogen left at the end of the reaction. The table shows the volume of ammonia (in m^3) produced at various temperatures and pressures when 1 m^3 of nitrogen is reacted with 3 m^3 of hydrogen.

Temperature	*Pressure (in atmospheres)*				
(°C)	25	50	100	200	400
100	1.91	1.95	1.96	1.98	1.99
200	1.56	1.69	1.80	1.88	1.94
300	0.86	1.13	1.38	1.60	1.77
400	0.32	0.53	0.81	1.12	1.43
500	0.11	0.21	0.38	0.62	0.97

From the above table it can be concluded that the amount of ammonia produced increases as

A. the temperature increases and the pressure decreases.
B. the temperature increases and the pressure increases.
C. the temperature decreases and the pressure increases.
D. the temperature decreases and the pressure decreases.

3

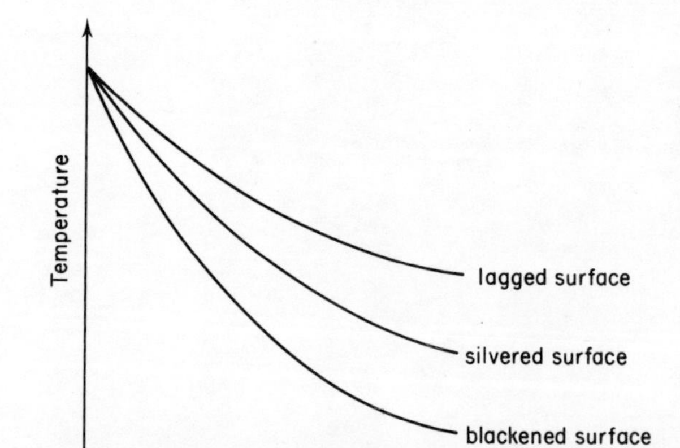

In an experiment to investigate the insulating properties of different surfaces, some cooling curves were plotted. The diagram shows the curves for three types of surface. Using only this information, which conclusion is most satisfactory?

A. The silvered can is better at keeping in heat energy than the lagged can.
B. The blackened can loses its heat energy the most slowly.
C. The lagging prevents the surface losing any heat energy.
D. The type of surface will affect a container's ability to keep in heat energy.

4

A detective was investigating a car crash and he found some headlight glass. One piece was missing, but he found a heap of glass and broken headlights in a suspect's garage. Which piece of broken glass would give him a clue? Which one fits?

Diagram 1

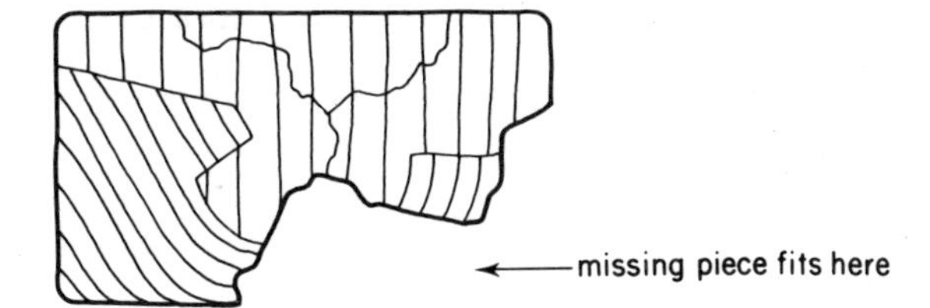

Headlight pieces found at the crash site

Diagram 2

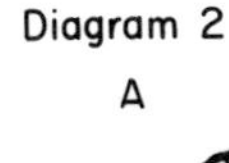
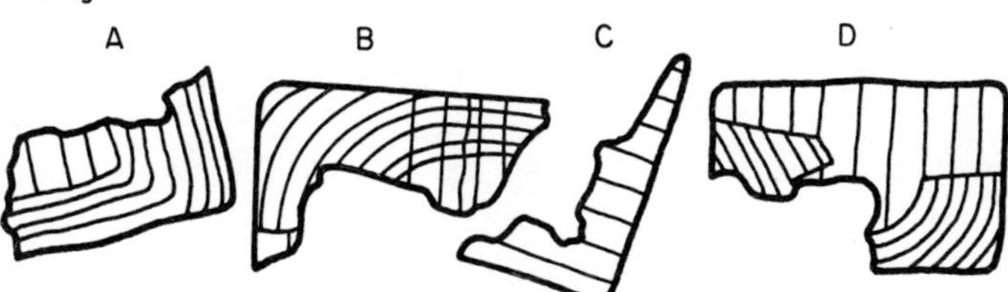

Headlight pieces found in the garage

5

The graph shows how the pHs of two rivers change over a year. One is in Sweden and one is in Canada. In both cases the amount of water in the river and the level of acidity is affected by melting snow.

There is an important pattern here. Which of these statements describes it?

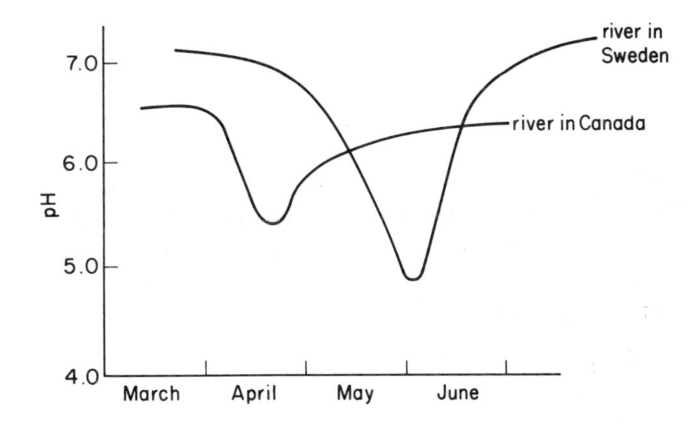

A. The acid levels show a sudden increase for about twenty days in the year and then go back to normal.
B. The acid levels show a sudden decrease for about twenty days in the year and then go back to normal.
C. The acid level in the Canadian river is always lower than in the Swedish one.
D. The acid level in the Swedish river is always lower than in the Canadian one.

6

A pupil did a series of experiments to find out about catalysts. Catalysts speed up chemical reactions. The results are shown in the diagram. Using only the above information, which of the following is the best conclusion about catalysts?

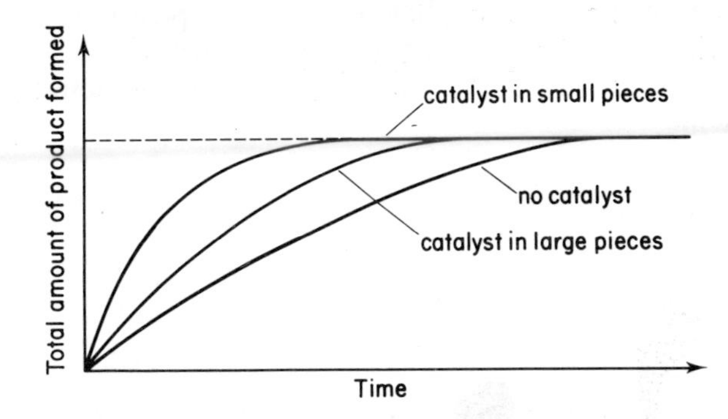

A. They produce more product from the same amount of reactants.
B. They remain unchanged at the end of the reaction.
C. They alter the rate of reaction and the products formed.
D. The reaction goes faster when the catalyst particles are smaller.

*7

Solder is a mixture of lead and tin but the percentages of lead and tin can vary. Different kinds of solder were melted and then allowed to go solid. The data in the table were obtained by plotting cooling curves as the solders solidified.

Solder	*Composition % lead*	*% tin*	*Temperature at which solidification begins* (°C)	*Temperature at which solidification ends* (°C)
W	67	33	265	183
X	50	50	220	183
Y	38	62	183	183
Z	20	80	200	183

From the above data it can be concluded that all solders

A. start to solidify at a constant temperature.
B. will start to melt at the same temperature.
C. contain the same amounts of lead and tin.
D. will become completely liquid at the same temperature.

8

One morning the pattern in diagram 1 is seen on the laboratory window. It has been caused by a bird. Which of the diagrams below shows the possible position and direction of flight of the bird when it produced the dirt? (Assume that there was no wind blowing.)

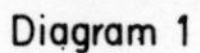

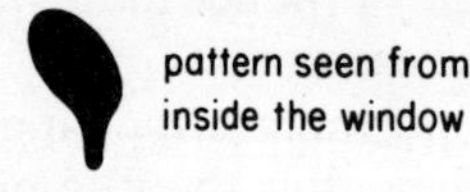

Diagram 2

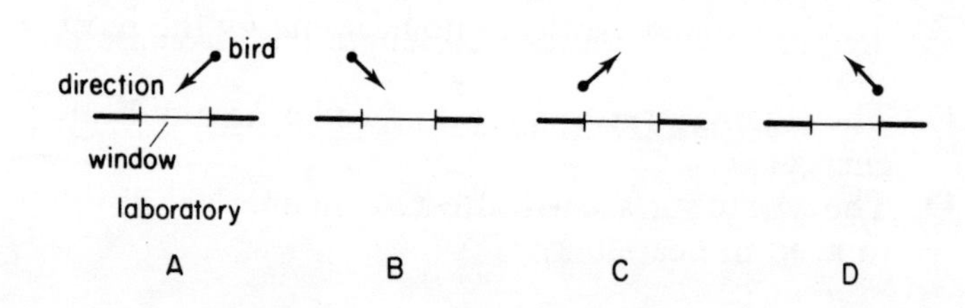

*9

A new element called polaikium has been discovered. It has the following properties:

(i) it is solid at room temperature,
(ii) it is a good conductor of electricity and heat,
(iii) many of its compounds are coloured,
(iv) it does not react with dilute hydrochloric acid.

The element's properties are similar to those of

A. aluminium.
B. carbon.
C. copper.
D. lead.

10

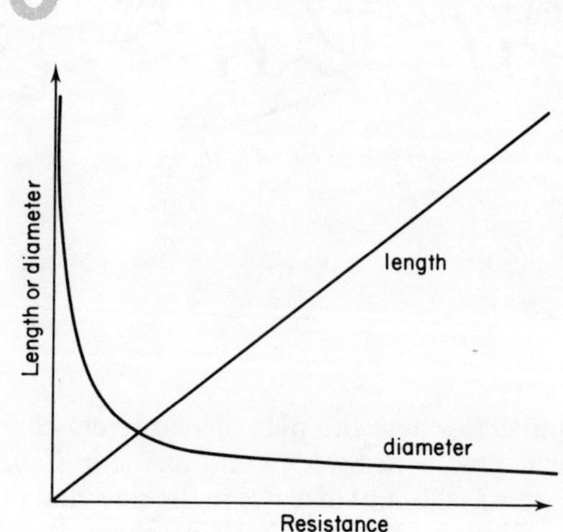

The graphs show how the resistance of a wire varies with its length and with its diameter. The resistance of the wire

A. increases with increase in length and increase in diameter.
B. decreases with increase in length and decrease in diameter.
C. increases with increase in length and decrease in diameter.
D. decreases with increase in length and decrease in diameter.

11

(a) Fish have nostrils to detect smells. This is important to fish which live where it is too dark to see food clearly. Birds and reptiles have dry, horny nostrils. They rely on sight to catch their food. Dogs can detect tiny traces of chemicals (drugs and explosives). What is the pattern?

A. Only land animals have a good sense of smell.
B. All animals can smell things equally well.
C. Sense of smell is best in animals which use it to catch food.
D. Smells can only be detected in air.

(b) There is a common theory about smell. All scientists believe that chemicals have different shapes. Some scientists believe that the lining of the nose has special receptor pits or holes. These pits also have special shapes. These chemicals and pits fit like keys to locks.

Look at the chemicals below and say which receptor pit each chemical would fit. One chemical does not have a matching pit.

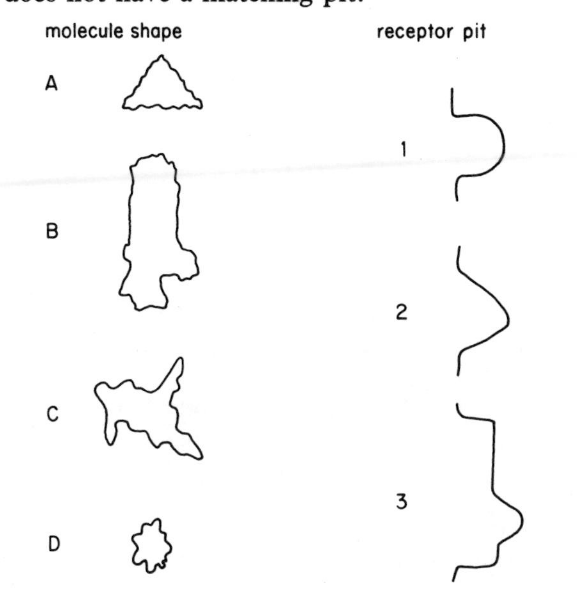

12

The table shows the percentage of people sensitive to smells. Each person stood 10 m away from each smell. They tested three age groups.

Age	*Sweaty socks*	*Sour milk*	*Frying bacon*	*Coffee*	*Curry*	*Onions*	*Perfume*
11–20	100	60	100	90	95	95	80
21–40	100	70	100	80	85	100	90
41–60	100	30	80	40	50	70	40

(a) What could be smelled by *all* age groups? (*1*)
(b) What pattern in the ability to smell do you notice in the table? (*2*)

13

People often try to forge other people's writing and signatures. Here are two pieces of writing.

(i) Well John there are many dividends to be earned if we invest this money
Jill.

(ii) Well John there are many dividends to be earned if we invest this money
Jill.

(a) Write down three clear similarities in the two examples of handwriting shown. (You can draw particular letters etc. if it helps.) (*3*)
(b) Write down three clear differences in the writing. (*3*)
(c) Do you think that B was written by the same person as A? Give one reason. (*2*)

14

If a tree trunk is cut across we see growth rings. This tree is 4 years old. It has 4 growth rings.

Diagram 1

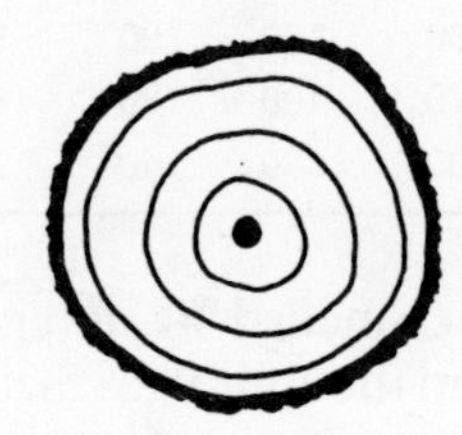

The trees below were planted at different times in the same garden. The drawings below show the growth rings seen when the trees were cut down. What pattern do you see linking the heights of the trees to the rings in the trunk?

Diagram 2

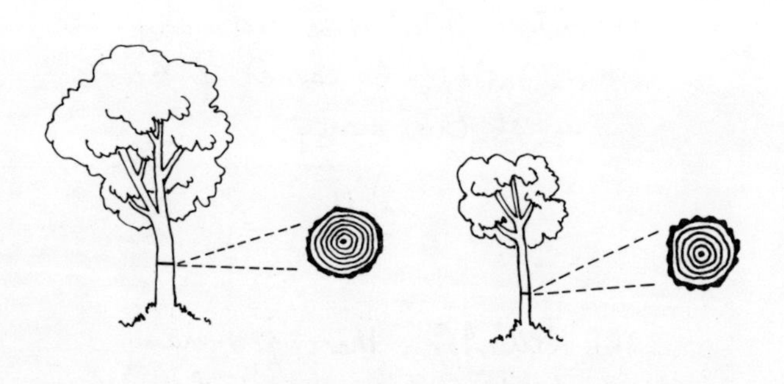

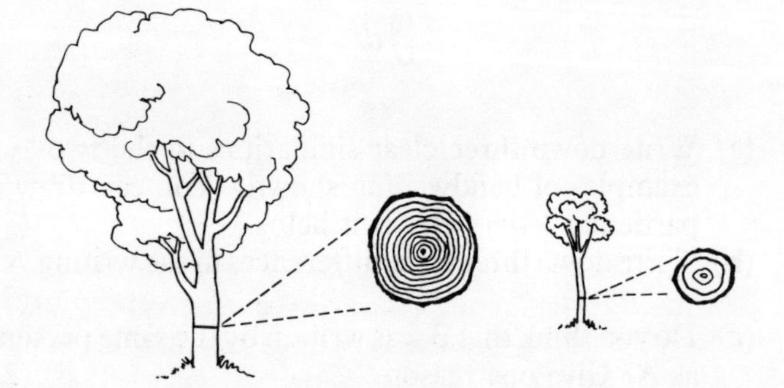

15

Some lenses magnify things. The diagrams below show four lenses (as seen from the side). Below each lens is a cross. Its size is roughly as it appears when you look through the lens.

Diagram 1

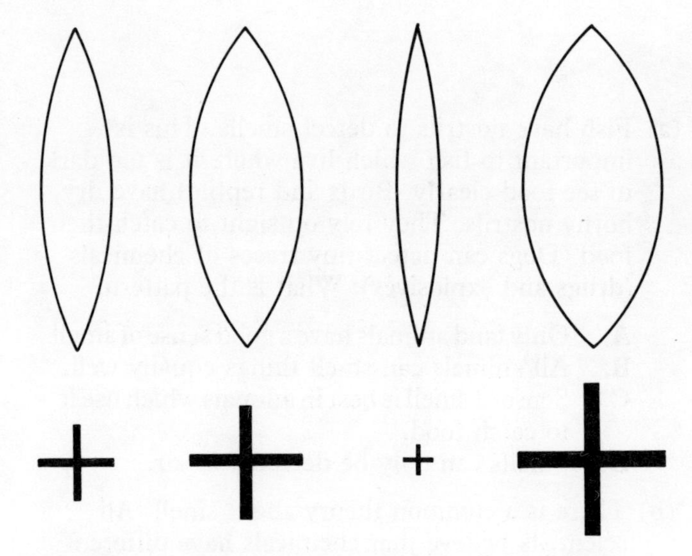

(a) What pattern do you see linking the width (fatness) of the lenses to the size of the cross? (3)

Diagram 2

(b) You are looking through this lens. Draw a cross the size you would expect it to appear if you look at it through the lens. (*1*)

(c) The size of the cross depends on where the lens is held, but you can use the data below to plot a graph and get a fairly accurate answer to part (b):

width of lens (mm)	0	5	10	15	20
width of cross (mm)	3	7.5	11.7	15	18

16

Some soaps are very good at removing certain marks. Repeated washing can also produce results. Copy out the table.

	Percentage of ink left after each time material is washed				
Brand of soap	*Once*	*Twice*	*3 times*	*4 times*	*5 times*
Luxury	90	80	70	60	
Gentle	80	60	40	20	
Soapy	95	90	86	83	
Pearly	100	100	100	100	
Smoothy	75	50	25	0	

(a) Look carefully at the table. Fill in the last column. *(3)*

(b) What general pattern do you notice about the effects of washing? *(1)*

17

The diagram below shows the demand for electricity during typical summer and winter days.

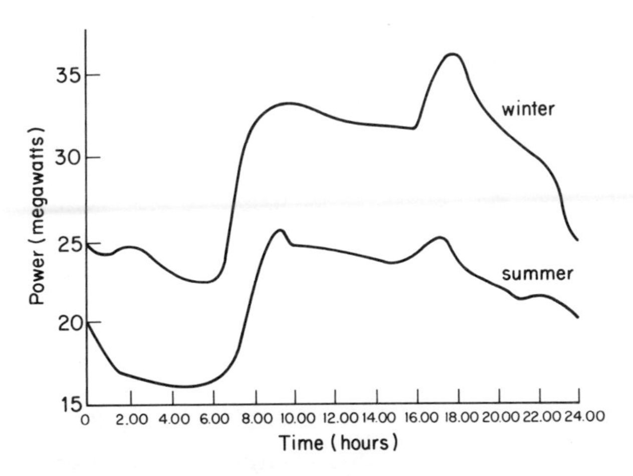

(a) Is there any similarity between the demands of the summer and winter day? *(3)*

(b) Is there any difference in the demand for a winter day compared with a summer day? *(1)*

(c) Explain the rapid rise in demand during the morning. *(2)*

18

For a project on radiators, some pupils wanted to find out the effect of colour and surface area (shape) on heat loss. They used containers which had different shapes and colours. The volumes of the containers were the same. They filled them with boiling water and left them for 30 mins. They measured the drop in temperature.

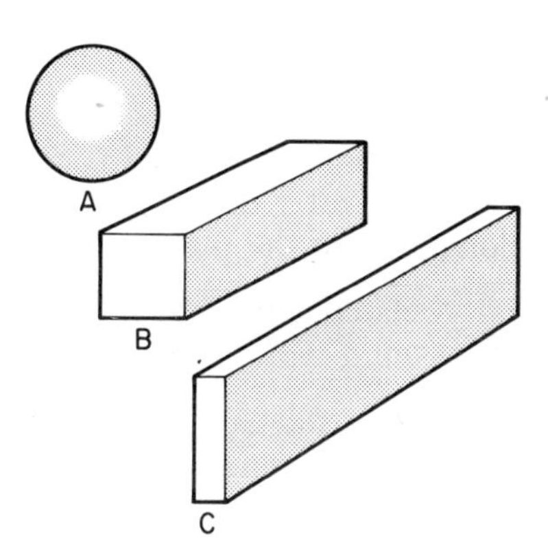

Colour	*Shape*	*Heat loss*
black	A	35°C
black	B	54°C
black	C	60°C
white	A	5°C
white	B	15°C
white	C	40°C

(a) Is white better than black at losing heat? Explain your answer. (2)

(b) Is shape important in heat loss? Explain your answer. (2)

(c) Describe the kind of radiator which would be best for warming a room. (2)

19

Baby mammals are fed on milk for a few weeks or months. The table shows the average length of the tail of baby mice after two weeks. The groups of mice were fed on different amounts of milk.

Milk per day (ml)	3	3.5	4	4.5	5	5.5
Av. tail length (cm)	5	6	7	8	9	10

(a) What pattern do you see linking tail length to amount of milk? *(2)*
(b) Predict (guess) the average length if some are fed 6.5 ml/day. *(1)*

*20 Plastic properties

Many plastics contain a plasticizer in addition to the polymer and this will alter the tensile strength and percentage elongation of the plastic under stress.

Tensile strength: related to the maximum load that the material can withstand without breaking.

Percentage elongation: related to the maximum increase in length a plastic can be stretched to without breaking.

Amount of plasticizer	*Tensile strength*	*Percentage elongation*
none	52–58	2–40
small amount	28–42	200–250
large amount	14–21	350–450

(a) The data in the table suggest that:
a tensile strength of 14 gives a percentage elongation of 450,
a tensile strength of 21 gives a percentage elongation of 350.
(i) Plot six points on a graph of tensile strength against percentage elongation. Draw the line of best fit. *(3)*
(ii) Label sections of the graph to show how much plasticizer has been added. *(2)*
(iii) Estimate the percentage elongation when the tensile strength falls to zero. *(1)*
(b) Polyvinyl chloride (PVC) has many uses. These often depend upon the amount of plasticizer. How much plasticizer would you expect to find in PVC used for
(i) drain pipes, *(1)*
(ii) plastic disposable cups, *(1)*
(iii) waterproof clothing? *(1)*
(c) Most plasticizers are non-poisonous liquids with high boiling points and low vapour pressures. Suggest reasons for these properties. *(4)*

*21 Group 2 in water

The table below shows how some metals in the same group of the periodic table react with water.

Metal	*Atomic number*	*Reaction with water*
beryllium	4	only reacts with steam at very high temperatures
magnesium	12	reacts with steam
calcium	20	fizzes slowly with cold water
strontium	38	fizzes fairly quickly with cold water
barium	56	reacts very rapidly with cold water

(a) What is the pattern between atomic number and the rate of reaction with water? *(5)*
(b) Suggest how radium (atomic number 88) would react with water. *(2)*
(c) How do you think
(i) beryllium, *(2)*
(ii) barium, *(2)*
would react with dilute hydrochloric acid?

22 Vive la différence!

The data below show the monthly variation in deaths for England and Wales and for France. The figures are based on an average of 100 deaths per month.

Time of year	*England and Wales*	*France*
Summer (average)	89.5	94.5
November	97.0	99.0
December	110.0	108.0
January	116.0	108.0
February	124.0	106.0
March	114.0	108.0
April	104.0	103.0

(Source: *New Scientist*, No. 1537, 4th December 1986, p.23.)

(a) (i) Is there any pattern in the above statistics for England and Wales? *(1)*
(ii) Give the evidence for your answer and suggest possible explanations. *(6)*
(b) Do the statistics for France follow a similar pattern to those for England and Wales? Give the evidence to support your answer. *(4)*

(c) Suggest reasons for any differences in the statistics for England and Wales and those for France. (*3*)
(d) Give possible reasons for the February figure for England and Wales. (*4*)
(e) Suggest ways in which the figures for England and Wales could be improved. (*5*)

(b) Shade the part that represents dangerous operating conditions. These are called deadman's curves. (*5*)
(c) (i) At what heights could a helicopter fly safely at 55 km/h? (*2*)
(ii) What is the minimum safe speed of a helicopter flying at a height of 50 metres? (*1*)

23 Helicopter safety

Helicopters sometimes have power failures. The rotor blades keep spinning by themselves as the helicopter falls out of the sky. Sometimes the spinning rotor blades slow down the fall enough for a safe landing. Sometimes they do not. It depends on the height of the helicopter when the power fails and on its forward velocity.

These helicopters crashed:

Height (m)	*Forward velocity* (km/h)
30	20
10	40
30	60
10	80
20	100
15	120
100	20
100	40
70	50
50	40

These helicopters landed safely after power failure:

Height (m)	*Forward velocity* (km/h)
5	50
10	60
30	70
60	70
30	80
100	50
80	100
30	120
60	120
25	100

(a) Make a pattern on graph paper showing operating conditions up to a height of 100 m and a speed of 120 km/h. (*5*)

24

Study the table.

Cigarette brand	*Nicotine* (μg)	*No. smoked per day*	*No. deaths per week*
U	80	10	100
V	75	15	110
W	90	20	125
X	85	25	125
Y	150	30	200
Z	180	30	250

Is it the nicotine or the number smoked or both which affects the number of deaths per week? Explain your reasoning.

25

A girl did an experiment to find out how the volume of a gas depends on pressure if temperature is constant. The following results were obtained.

Pressure (kPa)	70	80	90	100	110	120	130
Volume (dm^3)	14.30	12.50	11.10	10.00	9.09	8.33	7.69
1/*volume* (dm^{-3})	0.07	0.08	0.09	0.10	0.11	0.12	0.13

(a) Plot graphs of pressure (vertical axis) against
(i) volume (horizontal axis) (*4*)
(ii) 1/volume (horizontal axis) (*4*)
(b) (i) What conclusion can you come to from the pressure against volume graph? (*1*)
(ii) Give the relationship between pressure and volume. (*2*)

26 Budgeting metals

Metal	Production (tonne/year)	Price per tonne(£)	Percentage abundance in Earth's crust	No. of years reserve
aluminium	12700	750	8.1	257
chromium	6000	3700	0.01	110
copper	8000	1000	0.0055	42
gold	1	8600000	0.0000004	26
iron	400000	130	5.0	195
zinc	6000	500	0.007	27

Using the above data, answer these questions:

(a) (i) What is the pattern linking the annual production and the price per tonne? *(1)*
(ii) Which metal does not fit this pattern? *(1)*
(b) Which pattern links the amount in reserve and the percentage abundance? *(1)*
(c) Suggest why there is no clear pattern between percentage abundance and
(i) annual production,
(ii) price per tonne. *(2)*
(d) For copper and gold, calculate the total amount of reserves. *(2)*

27 Solar furnace

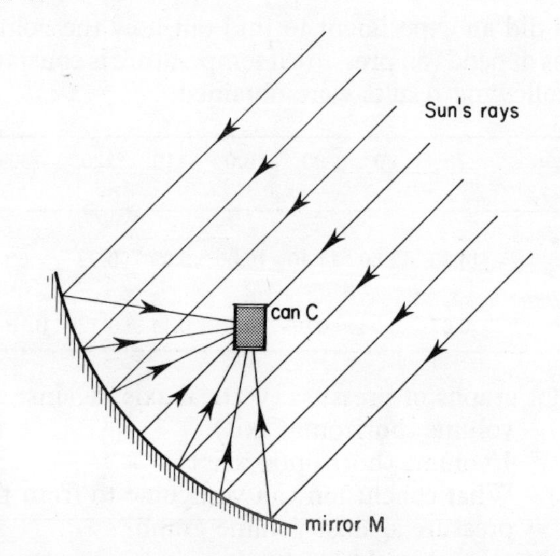

The diagram shows a solar furnace. Heat from the sun can be focused on the can C by mirror M.

Diameter of mirror M (m)	Maximum rates at which heat is received by can C (watt)
0.50	1060
0.75	2386
1.00	4240
1.50	9544

(a) How many times is the rate of heating increased if a 1.50 m mirror is used instead of a 0.75 m mirror? *(2)*
(b) What would be the rate at which energy is received by the can from a mirror of 2.0 m diameter? *(2)*

28 Gestation patterns

The time between fertilization and birth is called the gestation period. For a woman it is about nine months (282 days). The length of the gestation period depends on the size of the animal. It also depends on the way the mother looks after her young and the environment. The table gives the gestation period for a number of animals. Except for the whale, all these animals live on land.

Animal	Gestation period (days)
whale (8 m long)	350
elephant (3 m tall, 4 m long)	625
giraffe (4 m tall, 2 m long)	450
horse (2.25 m tall)	355
cow (1.75 m tall)	282
kangaroo (2 m tall)	42
chimpanzee (1.5 m tall)	250
armadillo (75 cm long)	115
dingo (60 cm long)	65
opossum (50 cm long)	13
rabbit (30 cm long)	40
mole (18 cm long)	30
mouse (8 cm long)	20

(a) What is the main pattern? *(2)*
(b) The kangaroo is an exception to the main pattern. Young kangaroos climb from the mother's womb to a pouch where they live for some time as babies and continue to develop. Which two other animals are exceptions to the main pattern? Explain why each of them may be exceptional. *(4)*

(c) Two cats are seen mating on 1st February. Use the pattern to predict when kittens might be expected. *(2)*

29 Printing photos

Diagram 1

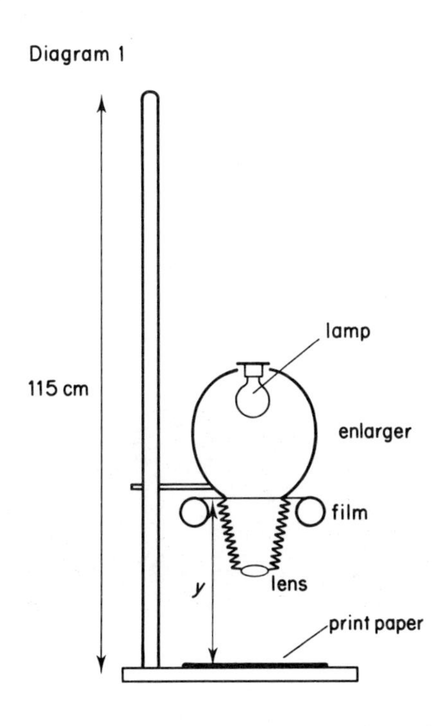

Miss Johnson develops films and prints photographs. She uses an enlarger. She wants to make a list to show what distance y should be used for particular enlargements. She finds that 30 cm gives × 2 enlargement and 40 cm gives × 3 enlargement. These are her results.

Distance y (cm)	*Enlargements*
30	×2
40	×3
50	×4
60	×5
70	×6

(a) What should distance y be for × 4½ enlargement? *(1)*

(b) What should distance y be for × 8 enlargement? *(1)*

(c) Look at diagram 1, which shows an enlarger. What do you think is the biggest enlargement that Miss Johnson could make? *(1)*

picture on film

enlargement x 3

30 Bridge tests

Karl wants to become an engineer. He tests a simple design for a bridge. He hangs a 1 kg mass from a strip of steel as shown in the diagram. The sag s seems to depend on the distance d. Karl makes some measurements and writes them down.

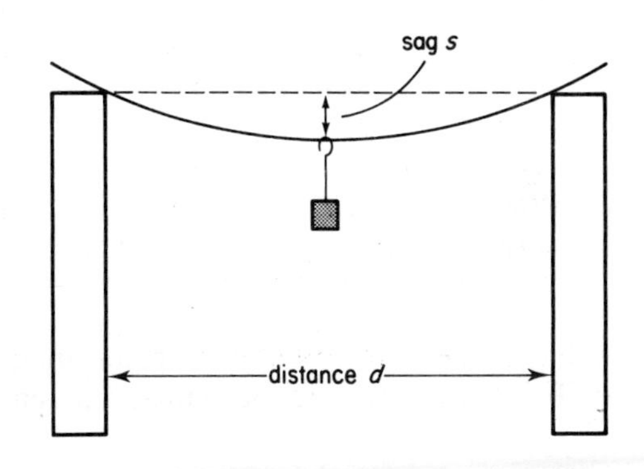

Distance d (cm)	*Sag s* (mm)
20	0.8
30	2.7
45	9.1
50	12.5
35	5.9
25	1.6
40	6.4
55	16.6

It is not easy to see whether Karl has made any mistakes in his measurements.

(a) Make another table. Put the information down in a more orderly way so that you can see whether or not there are mistakes. *(2)*
(b) One of Karl's measurements of sag is wrong. Without drawing a graph, find out which it is. *(2)*
(c) How did you work out your answer to (b)? *(3)*

31 Energy intakes

Human nutritional needs change as we grow older. Also, boys and girls need different amounts of carbohydrate for energy and protein for growth. Use the table below to answer the questions.

Recommended daily intake of protein and energy needs

	Male		Female	
Age (years)	Energy (MJ)	Protein (g)	Energy (MJ)	Protein (g)
<1	3.25	19	3.00	18
1	5.00	30	4.50	27
2	5.75	35	5.50	32
3–4	6.50	39	6.25	37
5–6	7.25	43	7.00	42
7–8	8.25	49	8.00	48
9–11	9.50	56	8.50	51
12–14	11.00	66	9.00	53
15–17	12.00	72	9.00	53
18–34	12.00	72	9.00	54

(a) (i) What is the pattern between recommended daily intake and age for both male and female? (4)
(ii) Give a reason for this pattern. (4)
(b) What is the pattern between the requirements for males and females? (2)
(c) The amount of energy, measured in kilojoules, provided by protein can be calculated using the formula

Energy = mass in grams × 17

(i) How much of the daily energy requirement for a young adult male comes from protein? (2)
(ii) Name the other sources of energy in the diet. *(1)*
(iii) How much energy does a one-year-old girl get from non-protein food? *(3)*
(d) During pregnancy the daily intake is 10.0 MJ of energy and 60 g of protein. During lactation (producing milk in the breasts) 11.5 MJ of energy and 69 g of protein are needed.
(i) What is the pattern compared with non-pregnant and non-lactating females? *(4)*
(ii) Suggest a reason for the pattern. *(2)*
(e) The recommended intake for a young male office worker who does little physical exercise is 10.5 MJ and 62 g of protein. How does this compare with the data given in the table? *(2)*

32 Fish scales

It is possible to work out the age of a fish by looking at its scales. This is a scale from a six-year-old fish.

Diagram 1

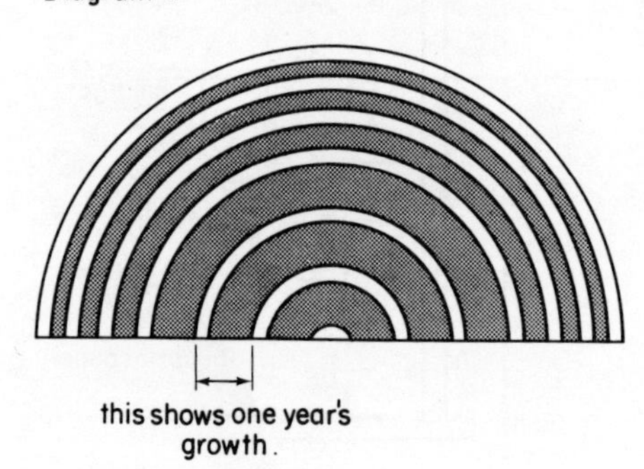

(a) (i) Describe the pattern of the fish's growth during the six years. *(3)*
(ii) What possible explanations are there for this pattern? *(2)*

This is a table of data from four different fish. Use it to answer the questions on the next page.

Diagram 2

scale pattern	age of fish (years)	length of fish (cm)	mass of fish (g)
		10	120
	5	12.5	140
		7.5	80
		15	150

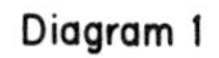

(b) (i) Copy out the table of results. Complete it by predicting the missing data. *(3)*
(ii) What is the pattern linking the number of rings to the length of fish? *(2)*
(iii) The entire fish is covered by scales, so what assumptions must you make when you compare scales from different fish? *(2)*
(iv) Does man show the same kinds of growth patterns? Give a reason for your answer. *(2)*

33

The table shows the density of a number of gases and their relative molar mass.

Gas	*Density* (kg/m^3)	*Relative molar mass*
carbon dioxide	1.98	44
carbon monoxide	1.26	28
hydrogen	0.09	2
methane	0.72	16
nitrogen	1.26	28
oxygen	1.44	32

(a) Which gas has half the molar mass of oxygen? *(1)*
(b) Compare the density of this gas with the density of oxygen. *(1)*
(c) Show how you use this pattern to find the density of ethane (relative molar mass 30). *(2)*

34 Stopping patterns

(a) The stopping distance of a car depends on the state of the driver and the state of the car. The pattern shown in Diagram 1 is for Mr Watson driving his own car.
(i) How does stopping distance depend on speed? *(2)*

Diagram 1

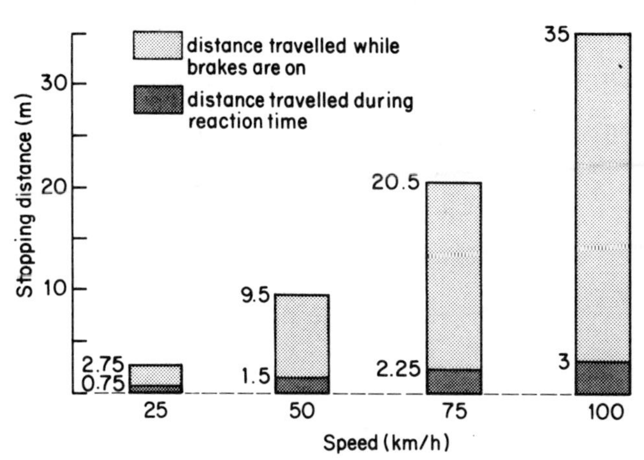

(ii) What is the pattern linking the distance travelled during reaction time and speed? *(2)*
(iii) What is the pattern linking the distance travelled while the brakes are on and speed? *(2)*
(iv) What would be the stopping distance if Mr Watson was driving at 125 km/h? *(3)*

(b) The pattern below is for Mr Watson driving a delivery van. What are the differences between the pattern for the delivery van and the pattern for the car? *(4)*

Diagram 2

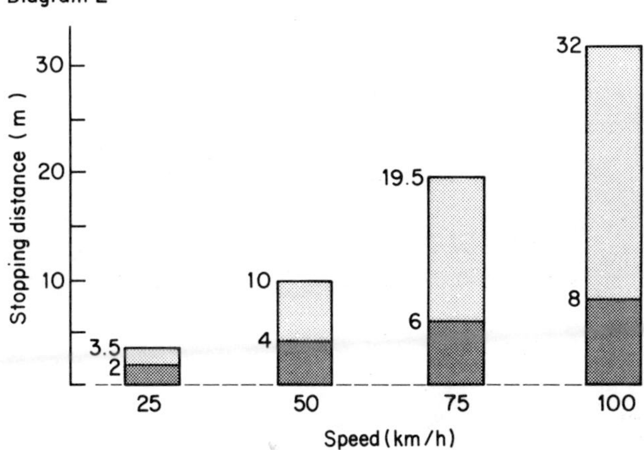

*35 Bimetallic strip

A bimetallic strip bends when its temperature changes. Bimetallic strips can be used as thermostats. Diagram 1 on the next page shows what happens to five bimetallic strips with change in temperature.

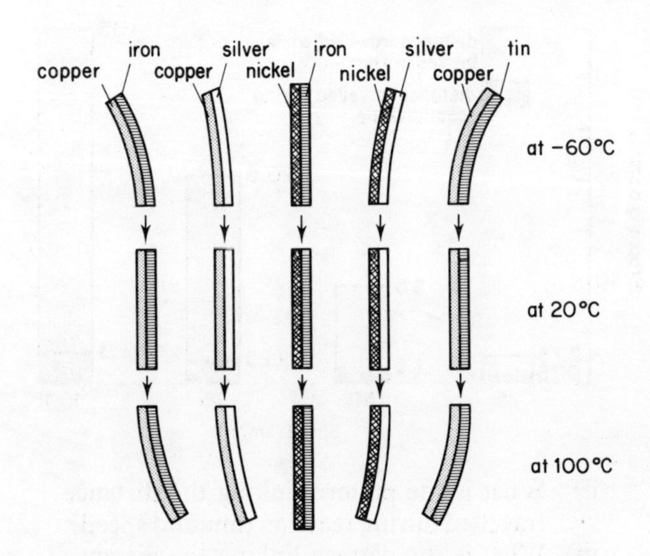

(a) (i) Which metal expands most when heated? (1)
(ii) Which two metals expand at the same rate? (2)
(iii) List the metals from the one which expands most to those which expand least. (3)

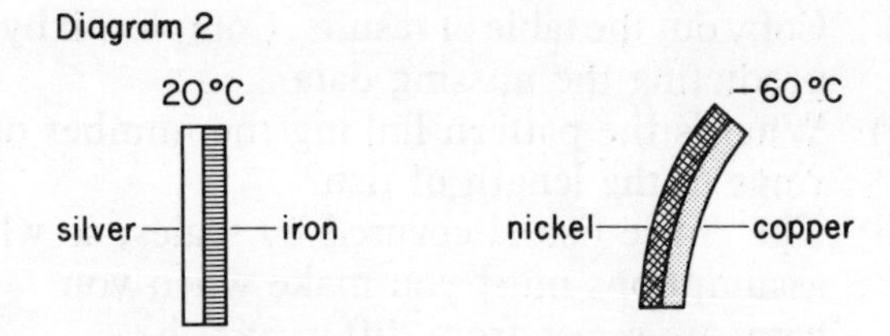

(b) Diagram 2 shows two bimetallic strips. The temperatures are stated. Draw diagrams to show the appearance of these bimetallic strips at 100°C. (4)

(c) Some thermostats are not very sensitive. If set at 20°C, they switch heating on at 18°C and switch it off at 22°C. This is partly because bimetallic strips do not bend much for a change of one or two degrees Celsius. Which pair of metals should be used to make a bimetallic strip in a very sensitive thermostat? (2)

4 Evaluation

1

Five people are looking at a plant in its pot.

Bill says, 'The roots must be damaged.'
Ashok says, 'The plant has been in the hot sun.'
Frank says, 'The plant stem is too weak.'
Steve says, 'The pot has no water in it.'
Susie says, 'The pot has a drooping plant in it.'

Who sticks closest to what they can see without jumping to conclusions?

A. Bill.
B. Ashok.
C. Frank.
D. Steve.
E. Susie.

2

Five people are looking at some rusty metal in a liquid.

Steve says, 'The liquid has made the metal rust.'
Karen says, 'The metal has been there for ages.'
Andy says, 'The liquid has some rusty metal in it.'
Husein says, 'The metal rusts quickly.'
Cynthia says, 'The water has made the metal rust.'

Who sticks closest to what they can see without jumping to conclusions?

A. Steve.
B. Karen.
C. Andy.
D. Husein.
E. Cynthia.

A

3

'Biopol' (polyhydroxybutyrate) is a new plastic which is being manufactured in Britain. It is biodegradable. It can be used to reduce litter problems because plastic bags and sweet wrappers made from it can be digested by bacteria. It can also be used for stitches after a surgical operation. The plastic stitches do not need to be removed. Bacteria in the body digest them. Which of the following would not be a suitable use for 'Biopol'?

A. making capsules containing medicine which can be swallowed.
B. making false teeth.
C. making pots for putting plants into the ground without disturbing the roots.
D. making packets for potato crisps.

4

The best powder used for fingerprint dusting on a window-frame which is painted cream is

A. talcum.
B. carbon.
C. iron filings.
D. sulphur.

5

Look at the picture of a bubble on the end of a straw. Which of the following can you definitely say is true by just looking at it?

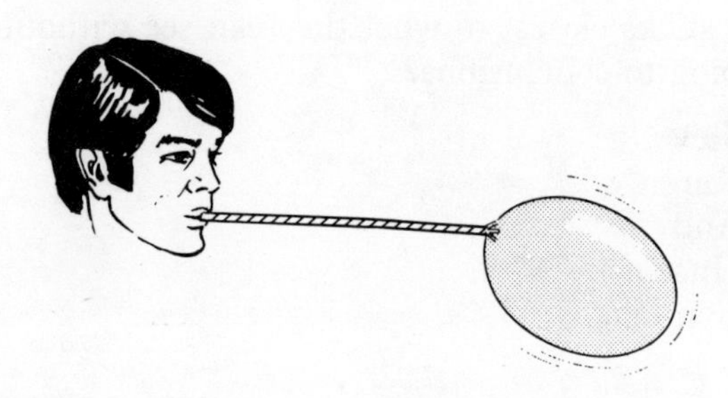

A. The person is blowing a bubble.
B. There is a bubble at the end of the straw.
C. The bubble is getting bigger.
D. The bubble is made of soap.
E. The bubble is about to burst.

*6

This is a diagram of a bicycle pump.

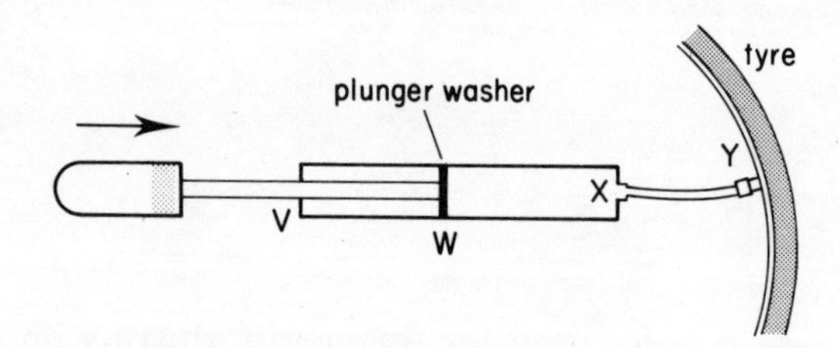

(a) When the pump is pushed in, where must there be an airtight seal if it is going to work?
(b) What happens if the pump is used and there is a leaky valve at Y?
 A. The handle jams.
 B. The plunger is forced back.
 C. The tyre deflates.
 D. Pressure rises at W.

7

In the spring, some birds fly north to Britain. In the autumn, they fly south again. This is called migration. Some small birds only migrate at night. If one of these birds is put in a cage where it cannot see local landmarks (church towers, hills etc.) but can see the night sky, it can pick the right direction for migration.

Some birds were hatched from eggs in laboratories. They never met other birds. They were never allowed to see the sky or stars. When released, they could find the right direction for migration. From this information, which of the following may be a correct conclusion?

A. Some birds have an inborn ability to recognize star patterns.
B. Birds find their direction by looking out for landmarks.
C. All migrating birds fly throughout the day and night.
D. Young birds learn migration patterns from their mothers and older birds.

8

The pie charts show the change in the use of energy sources in the USA from 1900 to 1950. The amount of coal used in 1950 was nearly as much as in 1900 because of the greater need for energy.

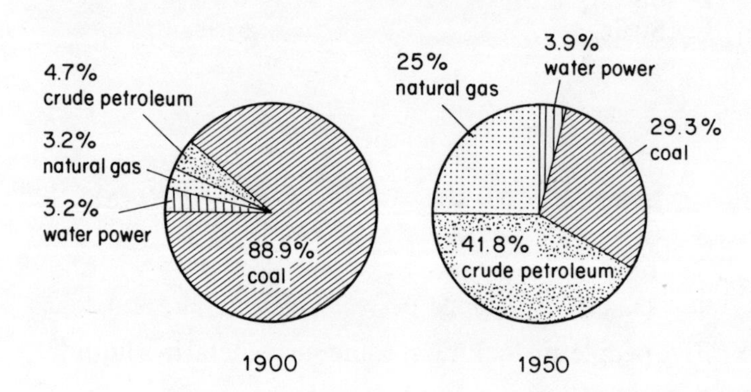

Consider these data and decide which of the following statements evaluates them correctly.

A. By 1900 most of the possible sites for hydroelectric power stations had been found and put to use.

B. The total energy used in 1950 was about three times that used in 1900.
C. There were more tonnes of coal used in 1900 than in 1950.
D. The use of crude petroleum had increased by 41.8% between 1900 and 1950.

*9

The map shows coalfields in China and Japan. Steam turbine electricity generating stations use a lot of coal and a lot of fresh water (not seawater). The places marked A, B, C and D are possible sites for steam turbine generating stations.

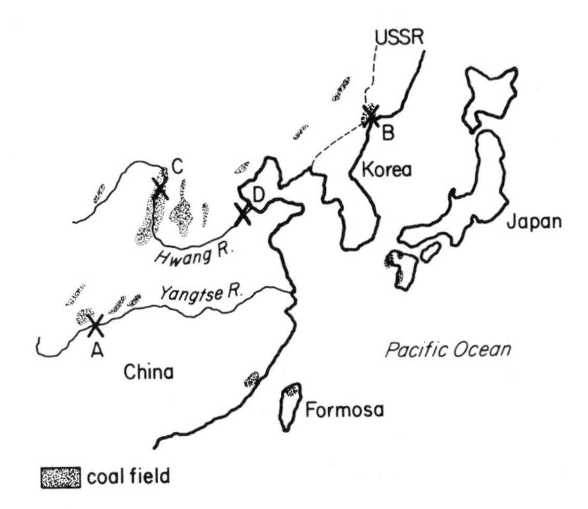

(a) From the information given, which would be the most suitable?
(b) Which of the pieces of information below would help you to make a better choice between the sites?
A. The climate (weather) at the site.
B. Whether or not there is valuable agricultural land on the side from which the wind usually blows. (Acid rain kills barley and may harm rice.)
C. Whether boats could get to the site.
D. The distance between the site and industrial regions.

*10

Some materials used to make furniture can give off deadly hydrogen cyanide fumes (formula HCN) when they smoulder. The material most likely to do this is

A. Polyethylene (polyethene): for example, coating on newspaper rack.

Diagram 1

$(-CH_2-CH_2-)_n$

B. Polyvinylchloride (polychloroethene), for example, easy chair covering.

Diagram 2

$(-CH_2-CH(Cl)-)_n$

C. Polypropylene (polypropene) for example, child's high chair.

Diagram 3

$(-CH_2-CH(CH_3)-)_n$

D. Polyurethane, for example, foam chair cushion.

Diagram 4

$(-(CH_2)_6-N(H)-C(=O)-O-(CH_2)_4-O-C(=O)-N(H)-)_n$

*11

Most soap powders contain fluorescent materials. They stay in some fabrics after washing. When these materials absorb energy from electromagnetic radiation, they re-emit the energy as light. The most effective wavelengths in electromagnetic radiation are about 0.37 μm. Which type of light will produce the brightest fluorescence from a fabric washed in soap powder?

A. Infrared (0.76 μm–5 μm).
B. Ultraviolet (0.04 μm–0.4 μm).
C. Red.
D. White.

12

Usually hot air rises and cold air sinks. This starts off a convection current. As air gets hot, it expands. So it becomes less dense than the rest of the air. Because it is less dense, it floats upwards. Gravity pulls the more dense, cooler air down to take the place of the hot air.

Diagram 1

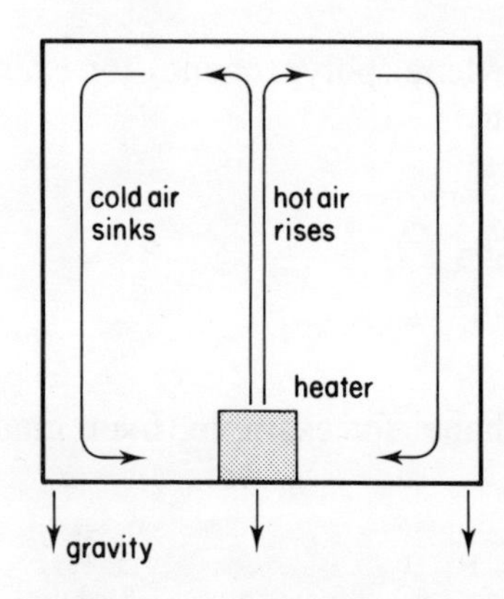

Diagram 2

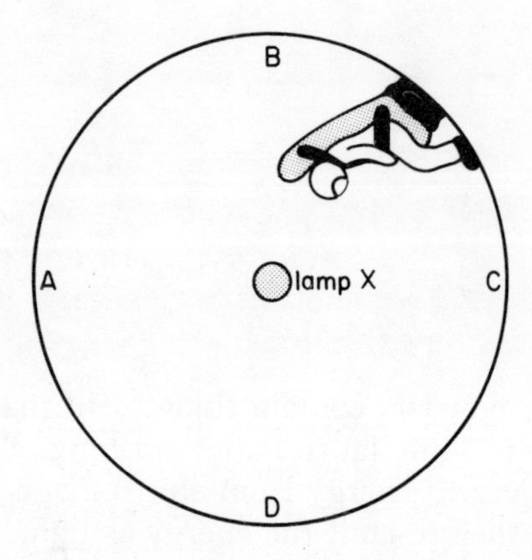

Diagram 2 shows the cabin inside a spaceship. The cabin is filled with air. There is no gravity in outer space. Hot air from the lamp X will

A. move to A.
B. move to B.
C. move to C.
D. move to D.
E. stay where it is.

13

We all receive nuclear radiation. It comes from a lot of different sources. The diagram shows how the radiation dose is made up for the average person. Which of these statements is a correct evaluation of these data?

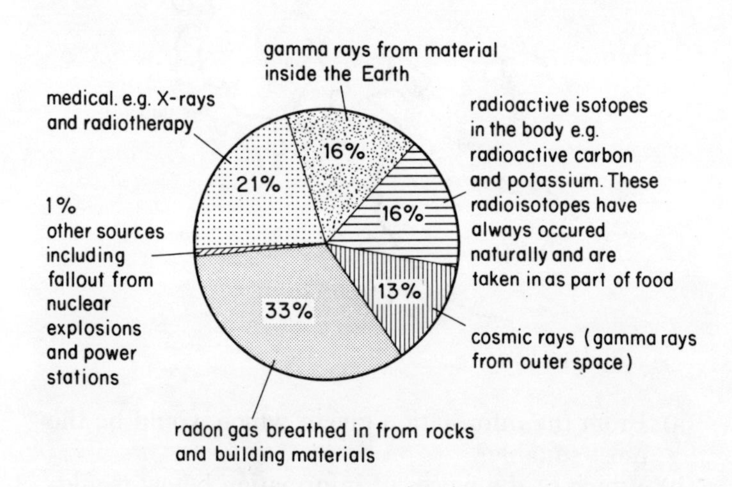

A. A worker in the nuclear industry is likely to get less than 1% of his total radiation dose from his place of work.
B. A hundred years ago, the average radiation dose was only about 22% less than it is today.
C. Even a hundred years ago, 33% of the average radiation dose was radon gas from rocks and building materials.
D. A person who never has X-rays or radiotherapy would still get 13% of his radiation dose from cosmic rays.

14

Read the passage below very carefully.

Siobhan lives on a farm and looks after the sheep. She has a well-trained and very faithful sheepdog. Shepherds control their dogs by whistling and calling to them. She was out one windy day trying to round up some sheep. Her dog was lying down close to the sheep. She whistled to her dog but it ignored her.

Write down three possible reasons to explain why this happened.

15

A pupil says, 'Professional microscopes are better than hand lenses.' Think of all the different things she might mean by this.

Some of these things could never be checked to find out if they are right or wrong. This could be because either they are just someone's opinion, or else because they are not clear. Choose one of the things she might mean that really could be checked, and write it down.

16

The diagrams show the results of a simple experiment to compare washing-up liquids to see how well they make a lather. Some liquid was put into a container with some water and shaken. The height of the lather was measured. Give two reasons why this was not a fair experiment.

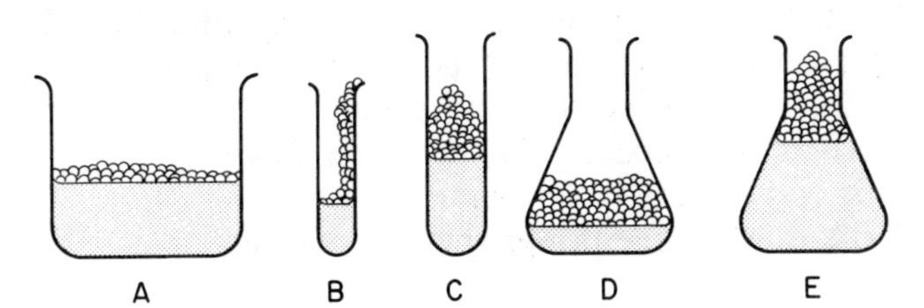

17

Friction is a force that stops things moving. It is harder to push something on sandpaper than on glass. This is because sandpaper is much rougher. Oil helps to reduce friction. This is why it is added to wheel axles, chains, cables, gears etc. on a bicycle.

Some pupils wanted to find which of two toy cars was the fastest. This is what they did.

Car A	*Car* B
(a) They oiled the wheel axles.	(a) 100 cm of card was used to make a steep slope with four books.
(b) 100 cm of card was used to make a shallow slope with two books.	(b) The car was released and the time taken to reach the bottom was measured.
(c) The car was pushed gently and the time taken to reach the bottom was measured.	

This was not a fair investigation at all. Write down three things they did wrong.

*18

There are two fields side by side. Barley was planted in both of them. The barley in field A is much bigger than in field B. Think of three reasons for this difference.

19

There are some factories that burn down because of accidents. Last year throughout the world, 6 icing sugar factories, 7 fertilizer factories and 4 flour mills burnt down after explosions.

(a) What have these factories in common that might have been a cause of the explosions? *(2)*

(b) Explain why some dusty factories spray water into the air or have extractor fans in certain areas in order to reduce fire hazards. *(2)*

20

Look at the picture and make a list of three things that could result in an accident.

21

Some flowers produce sweet nectar or have a pleasant smell to attract insects like bees to pollinate them. Other flowers also have bright colours and enticing shapes. There are a few unusual plants that have had to develop other methods of attracting pollinators. In South Africa there is a plant called the *Stapelia*. It aims to attract flies by:

(i) having petals which are brown, wrinkled and covered with hairs,
(ii) producing a horrible smell,
(iii) giving off heat.

(a) The petals are trying to imitate something which will attract flies. Suggest what they are trying to mimic. (*1*)
(b) Suggest what the plant might smell of. (*1*)
(c) What could the plant be mimicking by giving off heat? (*1*)
(d) Suggest what flies might do when they visit these flowers. (*1*)

*22

Look at the diagram of the car braking system. It is full of brake fluid which is shaded grey. Pressure is put on the fluid when the pedal is pressed and a piston is worked.

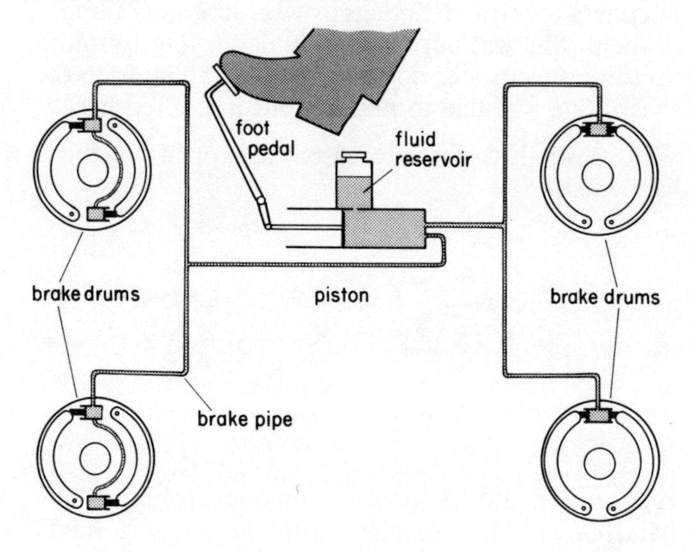

(a) If there was a small hole in the piping, what would happen to:
  (i) the pressure in the fluid, (*1*)
  (ii) the action of the brakes, (*1*)
  (iii) the action of the foot pedal? (*1*)
(b) How is the brake fluid topped up automatically? (*2*)

23

Some pupils wanted to see how easily a person's sense of taste could be confused by strong smells or a bad cold. This is the test they did.

(a) John, who had a bad cold was given a piece of food.
(b) He chewed it for 2 minutes and then had to identify the taste.
(c) This was repeated for several foods.
(d) Rajan, who did not have a cold was blindfolded.
(e) He was given a piece of food to chew for a minute and then asked to identify it.
(f) This was repeated for the other foods.

This was not a fair experiment.
Write down three things they did wrong.

*24

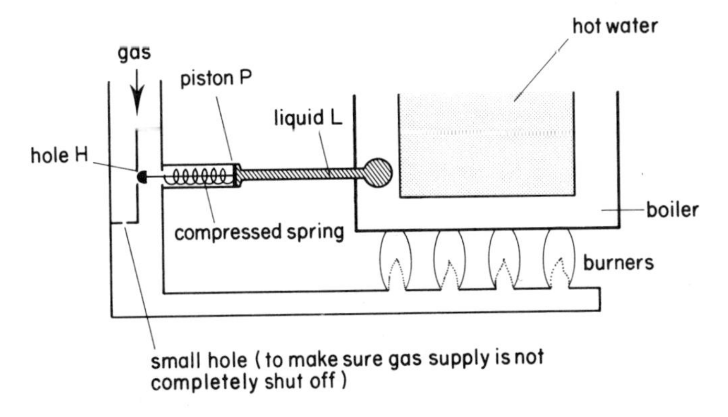

The diagram shows a device which could be used to control the supply of gas to a central heating boiler.

(a) What happens to the volume of liquid L when the boiler temperature increases? *(1)*
(b) What effect does this have on piston P? *(1)*
(c) What effect does this have on the flow of gas to the burner? *(1)*
(d) When the burner cools, what produces the force which opens hole H? *(1)*

25

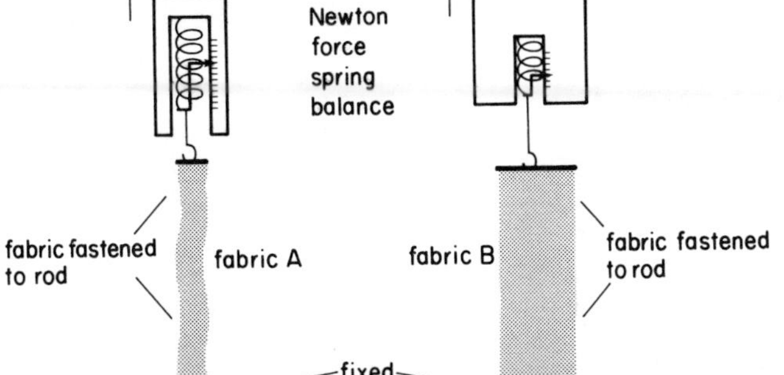

The diagram shows two sets of equipment. They were set up to test whether fabric A was stronger than fabric B. These two sets are not identical. This makes it an unfair test. Some of these differences are not important to the test. Write down two differences that do not matter.

*26

This question is about a house hot water system. Study the diagram and answer the questions.

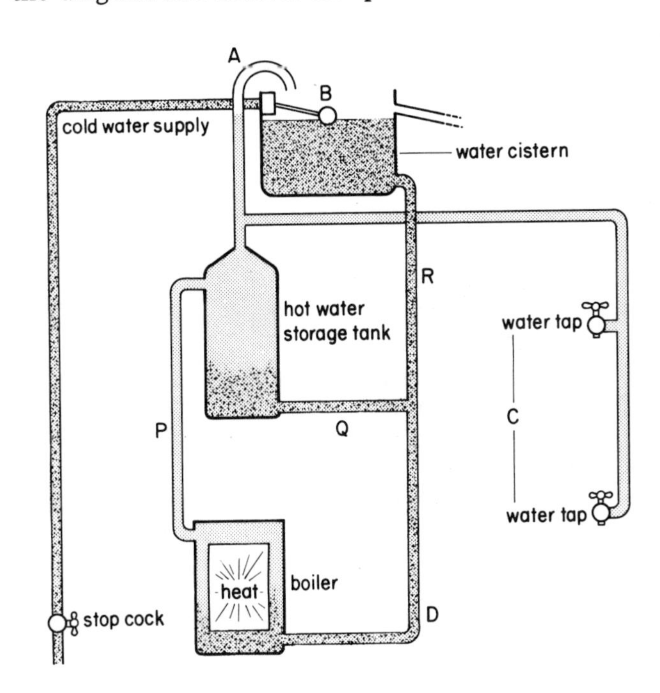

(a) Which way is the water moving at
  (i) P,
  (ii) Q,
  (iii) R? *(3)*
(b) Why is the pipe A necessary? *(2)*
(c) Will hot or cold water come from the taps labelled C? *(1)*
(d) Where is there an automatic tap? *(1)*
(e) Is the water at D hot or cold? *(1)*

27

Some pupils had been studying smoking in science. They knew that cigarette smoke gives off fumes, some of which are acid. They decided to compare two brands of cigarettes. Both had filter tips. They would test the acidity of the gases given off. This is what they did.

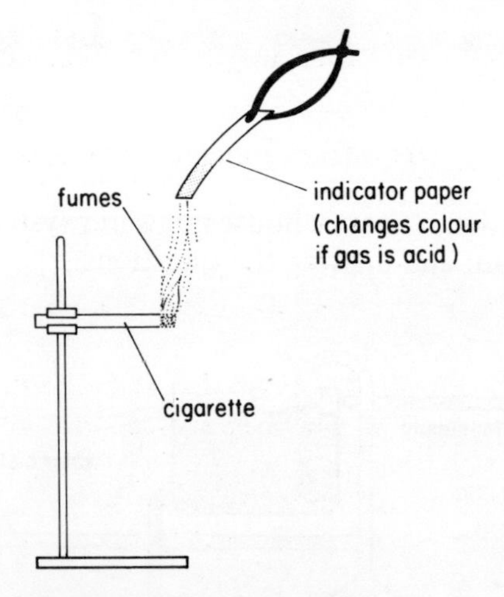

(a) A filter cigarette was set alight.
(b) A piece of damp indicator paper was held 15 cm above, in the fumes.
(c) They looked at the colour change after 20 seconds.
(d) The other cigarette was set alight.
(e) A piece of damp indicator paper was held in the fumes from time to time.
(f) The colour change was compared with that of the other cigarette.

This was not a fair experiment.
Write down three things that they did wrong.

28

Fred was asked to test three different concrete mixes for load bearing ability. He made three concrete beams, then he supported them on bricks and hit them hard with a large hammer. He used the same hammer for each beam. He counted the number of blows it took to break each beam. Then he recorded his results. They are set out below.

Experiment to test three concrete beams for load bearing
Apparatus

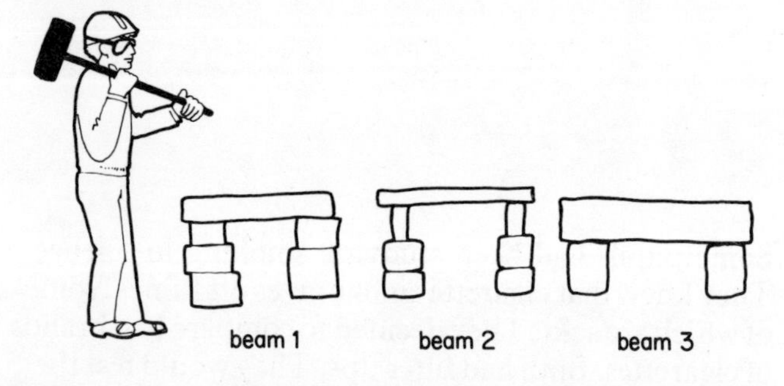

Bricks, large hammer, the three beams

Method
Each beam was placed across two piles of bricks. It was hit in the middle with the hammer. I counted the number of hammer blows it took to break each beam.

Results
Beam 1 One blow.
Beam 2 Three blows.
Beam 3 Six blows.

Conclusion
Beam 3 was the hardest to break. It must be made of the best concrete mix for load bearing.

This experiment is full of errors. List four and say how you would correct them.

29 Chip machine

This is a diagram of a chip making factory process. Use it to help you to answer the questions.

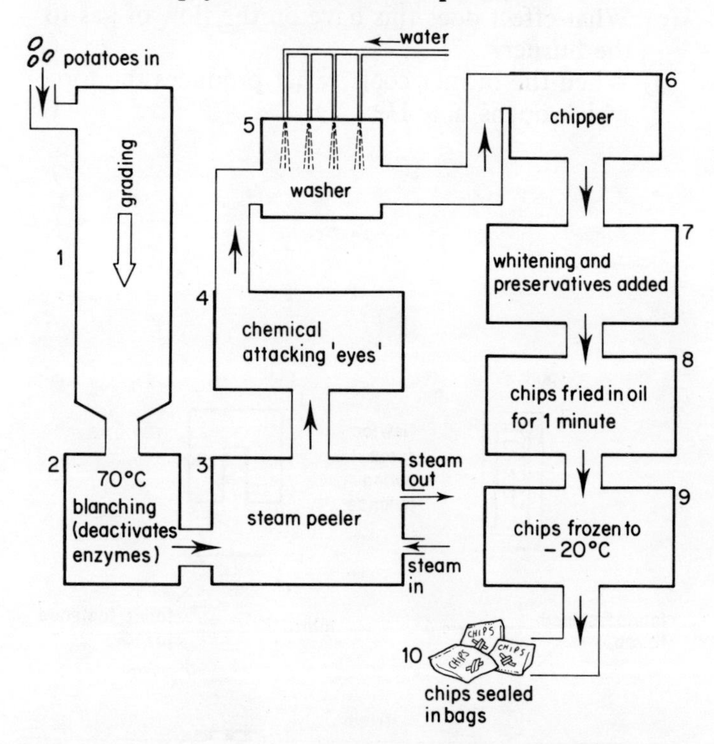

(a) Why would the potatoes have to be graded for size? *(2)*
(b) What happens to the potatoes at 70°C? *(1)*
(c) How long are the chips fried before freezing? *(1)*
(d) Why is step 5 essential? *(2)*
(e) Why must the chips be whitened? *(2)*
(f) What is the freezing temperature for the chips? *(1)*

*30 Glass stopper

Darren and Richard have just begun an experiment with a spring balance reading grams. They have a beaker of water and a glass stopper. Their problem is to find the volume of the glass stopper when there is no measuring cylinder.

This is what they did and said. Answer the questions as you go along.

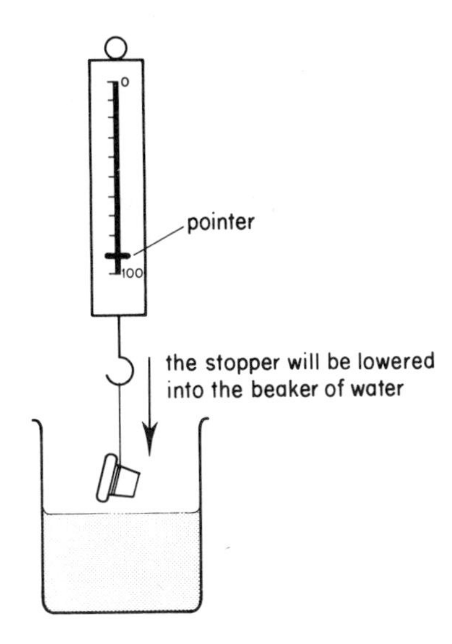

(a) D: The spring balance pointer will go up.
R: No it won't. It will go down.
D: I think it will go up.

(i) Who is right?
(ii) What should Darren have said about the pointer, to be more accurate? *(2)*

(b) R: The water level has gone up too.
D: So it has. That must be the weight of water displaced.
R: Don't be silly. You can't have a weight and a water level!

(i) What did Richard mean, when he said that you can't have a weight and a water level?
(ii) Could Darren have said, 'That must be the amount of water displaced'? *(2)*

(c) D: Let's work out the answer. The stopper in the water has pushed the level up . . . .
R: Yes, and it pushed the pointer up.
D: Then this water is holding up the stopper . . . .
R: Why isn't it floating then?
D: It doesn't have to. It can be held up a little bit. That's why it seems lighter. What does the pointer say?

(i) There is a false conclusion in this section. What is it? *(1)*
(ii) Is Darren's statement about the stopper seeming lighter an accurate statement? *(2)*

(d) R: The pointer says 50 g, so the stopper's volume is 50 g.
D: That doesn't seem right. It's too much.
R: But we started at 90 grams, so the volume must be 40 grams.
D: That seems better, 40 grams.
R: Volume isn't measured in grams.
D: It must be 40 cm^3 then.

(i) What did Richard forget at first? *(2)*
(ii) Explain whether Darren has made a good estimate of the volume. *(2)*
(iii) How can you change from measuring in grams to measuring in centimetres cubed? *(2)*
(iv) These boys should have planned their work first before starting on their practical assignment. Write out a plan for working out this problem. Say where you would have acted differently. *(4)*

*31 Snail race

Elaine and Kim were asked to say which of two snails was 'the best', by taking measurements. This is what they said. There are questions about what they are doing. Answer the questions as you read.

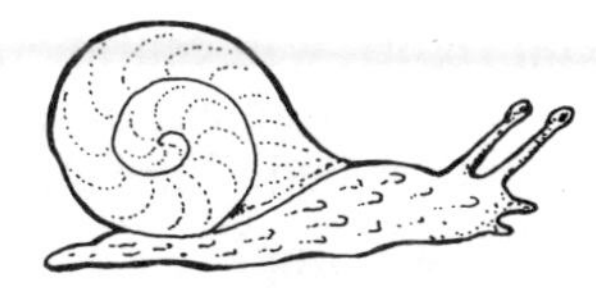

5.0 g
Elaine's snail

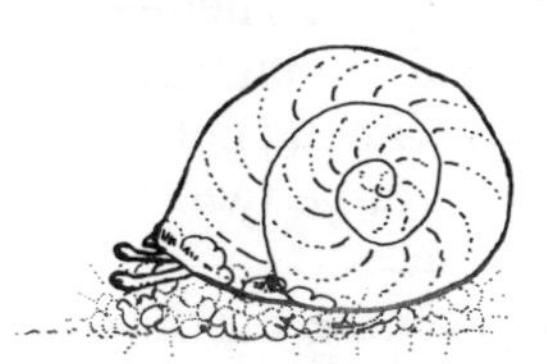

7.5 g
Kim's snail

(a) E: Mine's best, it's got yellow stripes on its shell.
K: Mine can blow bubbles.
E: What shall we measure?
K: Does yours blow bubbles? Poke it with a pencil.
E: No, it's cruel.
K: I will then. Look! Mine's blowing more bubbles . . .
E: How do you know? Mine's only just begun.
K: Right! Estimate from now for 1 minute.

(i) Explain how Elaine's observation about yellow stripes could lead to measurement. *(3)*

(ii) Snails blow bubbles when they are frightened. What is likely to happen after 1 minute? *(1)*

(b) E: Mine has the most bubbles. It's the winner!
K: Mine's not as slimy. It's looking at me! Oh, its eyes are up on its feelers so it can see better. Mine's got longer feelers, so it can see better.
E: OK! I've got one mark and you've got one mark. Let's race them to decide.
K: Good idea! Line them up by the ruler on the smooth table.
E: We should feed them on cabbage.
K: Right! Are we measuring how fast they go or how far?

(i) Explain whether Elaine is justified in claiming that her snail is the best bubbler? *(2)*

(ii) Would Kim's snail really be able to see better? Explain your answer. *(2)*

(c) K: We need a stopwatch.
E: What for? If we measure how far we don't need one.
K: Good! I can't work them anyway. We'll measure in centimetres because that will show the differences clearly. I need a centimetre ruler. Mine's only got inches.

(i) Do they need a stopwatch? Explain your answer. *(2)*

(ii) Is Kim right? Will centimetres give a more accurate measurement of difference in distance travelled? Explain your answer. *(2)*

(d) E: My snail's winning!
K: Mine's more steady.

(i) Elaine's snail did win. It travelled 50 cm. Kim's snail went more slowly and it travelled 33.5 cm. Explain whether this race was a fair test of speed or stamina. *(2)*

(ii) Elaine and Kim did not think of any variables to control. Name two that might affect their investigation. *(2)*

(iii) How could more measurements be taken from this investigation? *(2)*

(iv) How would you improve this investigation when it is your turn to do it? *(3)*

*32 Sweaty students

Some students wanted to find out how much different people sweat. The students weighed themselves and then went into a sauna (steam room) for 1 hour. They were re-weighed afterwards. The bigger the difference in masses the more sweat is made.

Alan, Ranjit, Tom and Winston went in the sauna on Tuesday night after having a drink together at a public house. Anne, Kathleen, Susan and Tara went in on Wednesday night, going straight from lectures. It was Susan's first time at the sauna and she took a large towel in to keep parts of her body covered up. Tom had to visit the toilet after being in the sauna for half an hour. But he made up time by staying longer at the end. The men stood up most of the time but the women sat or lay on the benches. Winston managed to do a hundred press-ups while in the sauna. None of them ate anything but Alan chewed gum most of the time.

Evaluate this experiment. Explain why it did not give a fair comparison of the sweating rates of the eight people.

33 Vitamin C

The data below shows the amount of vitamin C in milligrams (mg) left in 100 g of Brussels sprouts after boiling for the time given.

Boiling time (min)	0	5	10	15	20	25	30	35	40	45
Vitamin C (mg)	37	31	28	25	24	23	22	21	20	19

(a) (i) Plot a graph of milligrams of vitamin C (vertical axis) against boiling time. *(4)*

(ii) How does the amount of vitamin C vary with cooking time? *(3)*

(b) Vitamin C is fairly soluble in water. The data below give the percentage of vitamin C still in vegetables after cooking in various ways.

Vegetable	*Pressure cooking (steaming)*	*Water to cover*	*½ cup of water*	*Waterless cooking*
cabbage	75.5	44.3	57.4	68.4
carrots	79.1	63.1	75.1	72.5
peas	73.7	51.3	70.0	78.8
potatoes	57.3	41.0	48.4	79.4

(i) How does the amount of cooking water affect the loss of vitamin C when cooking peas? Is this true for all vegetables? *(3)*
(ii) In many cookery books it is suggested that the vegetable cooking water should be used for making the gravy. Why is this suggestion given? *(2)*
(iii) Suggest the best way to cook cabbage. Give a reason for your choice. *(3)*
(iv) Using the above information, why are jacket (baked) potatoes nutritionally better than boiled potatoes? *(2)*

*34 Growing things

Caterpillars are serious pests on arable farms (where crops are grown). They must be controlled but chemicals can affect the balance of nature. DDT is a chemical pesticide that stays inside animals. It affects their fertility and sometimes kills them. Use this diagram of a food web to help you to answer the questions.

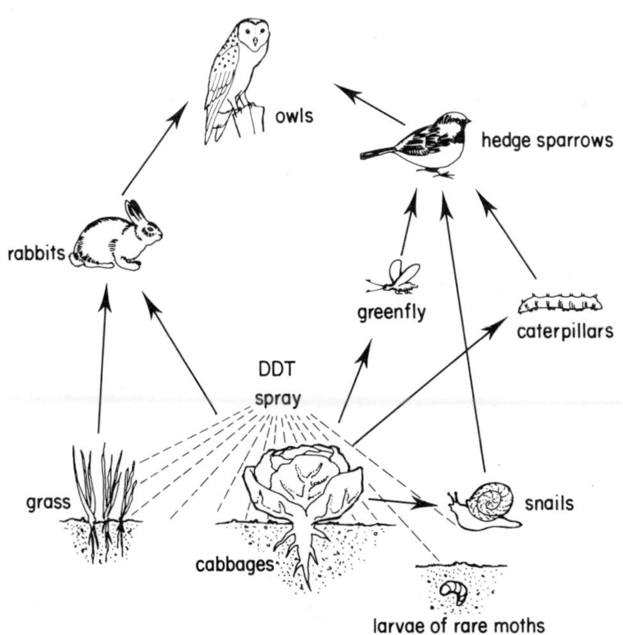

(a) Name two animals killed directly by spraying. *(2)*
(b) Which wild creature receives the highest dose of DDT? *(2)*
(c) Name two other ways in which DDT spraying can affect wildlife. *(2)*
(d) Ladybirds eat greenfly and other sprays are available. Give two pieces of advice to a farmer who wants to give up DDT, but still has pests on his cabbages. *(4)*

35 Magic bracelets

Read this advertisement and then answer the questions about it. Your answers should be about 3–5 lines each, not just 'yes' or 'no'.

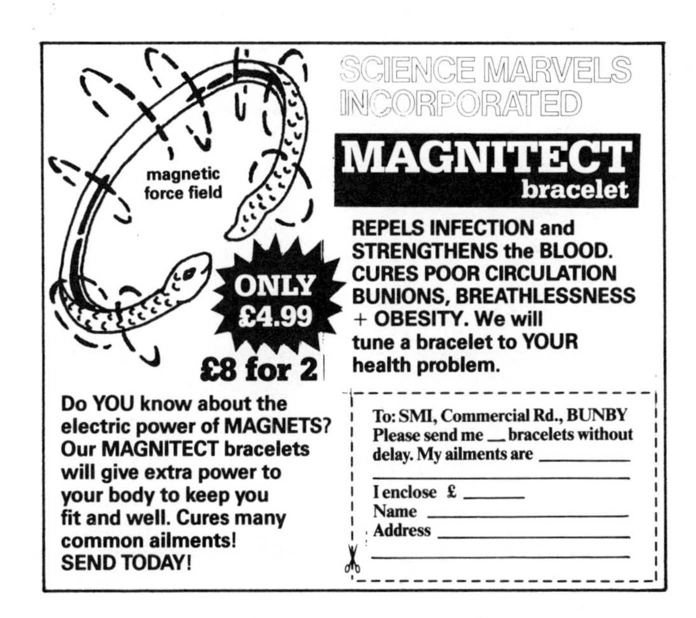

(a) Would it be reasonable to say that scientific research has produced 'marvels' in keeping our bodies fit and well? *(2)*
(b) (i) List four words used in the advertisement which have a precise meaning in science. *(4)*
(ii) The advertisement could be described as 'pseudoscientific'. What do you think this means? *(1)*
(iii) A boy who sends for a bracelet could be called 'scientistic'. What do you think this means? *(1)*
(c) What is the theory of the bracelet? How does Science Marvels Inc. explain how it works? *(3)*
(d) If a boy bought a bracelet, would his spots be likely to go away? *(2)*

*36 Tanks

Army tanks have to be able to go over icy land without slipping. They must also cross sandy beaches and muddy fields without sinking in. Tanks usually have caterpillar tracks with sharp spikes. Use the information given below to discuss the design of the tank in terms of pressure on different kinds of land surface.

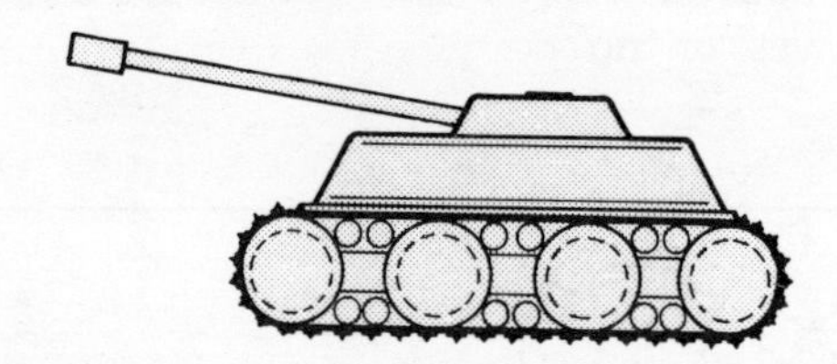

$$\text{Pressure} = \frac{\text{force}}{\text{area}}$$

The bigger the pressure, the more you sink into a surface.

*37 Window opener

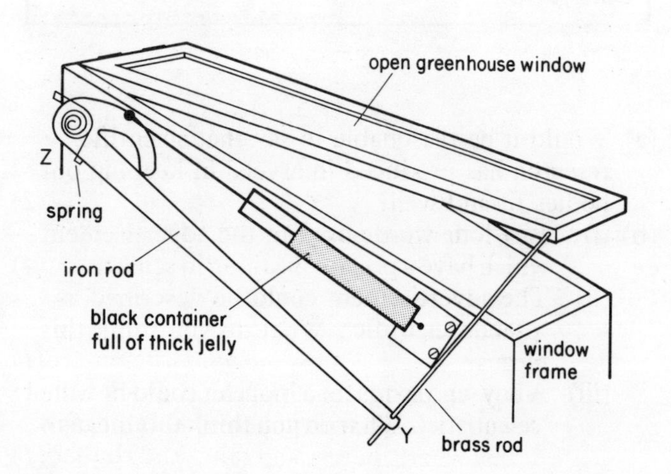

This is a diagram of an automatic window opener for greenhouses. It is fitted to a window frame. It works because jelly inside the container expands a great deal when the temperature rises. Use the diagram to help you to answer the questions.

(a) What is the purpose of this device? *(2)*
(b) Why is the expanding jelly inside a black container? *(2)*
(c) The window is shown in a half-open position. List the events that will:
  (i) make the window open more, *(2)*
  (ii) make the window close. *(2)*
(d) Why is there a spring at Z? *(1)*
(e) There is a wing nut at Y which allows the brass rod to be separated from the iron bar. Why is this useful? *(1)*
(f) The window turns on its hinges. These are pivots. How many other pivots are there in the diagram? *(1)*

*38 House insulation

There is a 5 cm space (cavity) between the inner and outer wall of a house. It stops dampness penetrating from the outside to the inside. Two neighbours are having a conversation about house insulation.

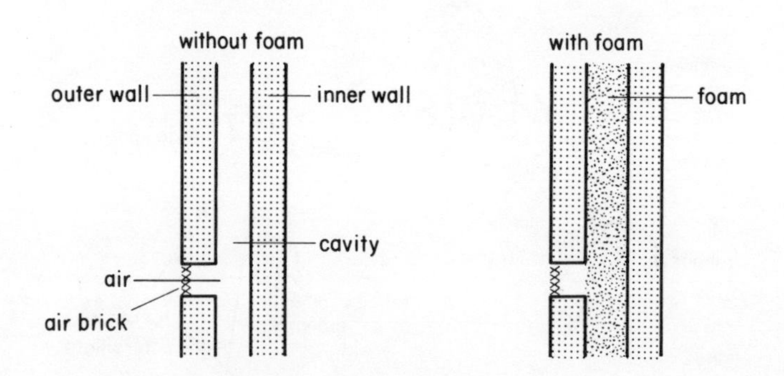

(a) Abdul: Hey Harry, why don't you have plastic foam pumped into your cavity like me? The plastic bubbles trap air and air is a bad conductor of heat.
Harry: So what? My cavity is full of air anyway. There's just as much air in my cavity as in yours. Why should foam keep the house any warmer?

Evaluate these statements. *(4)*

(b) Harry: What you need is double glazing. My double glazing cost £2000. But it will soon pay for itself. It cuts down heat loss by 20%. I used to pay £300 a year for heating.

Evaluate this statement. *(4)*

(c) Abdul: You have five times as much outside wall as window. So insulating walls is five times as important.

Harry: My double glazing cuts down heat loss by 20%. You are saying your wall insulation cuts down heat loss by 100%. What about your windows and roof?

Evaluate these statements. *(4)*

*39 Fresh air

In this question you will be asked to evaluate some statements. You will need to know about:

Photosynthesis, respiration, combustion, convection, diffusion, relative molecular mass, air pollution, and the effect of acid rain.

Some information

Element	*Relative atomic mass*
nitrogen	14
oxygen	16
carbon	12

Gas	*Formula*
nitrogen	N_2
oxygen	O_2
carbon monoxide	CO

Plants affected by acid rain:

Worst affected

decreasing effect of acid rain ↓

Lichens
Barley
Rhubarb
Lettuce
Sprouts
Cabbage
Potatoes
Onions
Celery

Least affected

Read this passage and evaluate the statements

Ethel and Maud live in Leeds. Ethel lives on the 10th floor of a block of flats in the city centre. Maud lives in a semi-detached house in a side street in the suburbs. They are on an old people's outing to Pateley Bridge in the countryside. It is a sunny day in June.

(a) Ethel: I like to get a nice breath of fresh air out in the countryside. There's more oxygen because of all those plants growing. Plants make oxygen you know.

Maud: I'm not so sure. Plants are alive aren't they? They've got to breathe like us. I reckon all those blades of grass breathe up a lot of oxygen: just like a bingo hall on a Friday night.

Evaluate these statements. *(4)*

(b) Ethel: Of course living up in my tower block there's a lot more oxygen about than down at ground level where you live.

Maud: I don't know about that. All those cars and lorries passing where you live use up a lot of oxygen. They make carbon monoxide too and that's poisonous. I bet it comes out of the exhaust pipe of a double decker bus and makes straight for your flat. Hot air: what can you expect? And besides, it's the molecules. Those carbon monoxide molecules are very light.

Evaluate these statements. *(6)*

(c) Ethel: With all those geranium plants you have in your bedroom, Maud, I'm surprised you're still alive. They make oxygen during the day but at night they use it all up.

Evaluate this statement. *(2)*

(d) Coach driver: Well here you are anyway. You're just above Pateley Bridge. Twenty five miles north-west of Leeds,with a good fresh wind from the south-east.

Maud: You're right, Ethel, it's a lot different from Leeds; none of that pollution. Look at that wall; covered with lichen. You can't grow that where there's acid rain. And rhubarb, Ethel. My Albert could never do anything with rhubarb in Leeds because of acid rain. Mind you he had some lovely onions. Onions are not so particular.

Evaluate these statements. *(5)*

*40 Indigestion remedies

Most of the drugs used to treat indigestion contain one of the following compounds:

(a) aluminium hydroxide,
(b) magnesium carbonate,
(c) magnesium hydroxide,
(d) sodium hydrogen carbonate.

(i) Suggest what is produced in the stomach to give the symptoms of indigestion. *(1)*
(ii) How do indigestion drugs help to relieve indigestion? *(2)*
(iii) Choose one of the compounds in the list and write a word equation for the reaction that occurs in the stomach. *(2)*

A *41 Castings

David Price makes washing machine parts. If aluminium is heated enough it melts. A washing machine part is made by pouring melted aluminium into a mould. When pouring starts, the temperature of the aluminium is about 30°C higher than its melting point.

(a) The size of the mould is bigger than the final size required for the part. Why is this? *(1)*
(b) Why is the temperature of the aluminium above its melting point when pouring starts? *(1)*
(c) David has to wear special heat-resistant clothing when pouring the aluminium. Why is this necessary? *(2)*
(d) The final volume of the part is 0.002 m^3. The density of aluminium is 2800 kg/m^3. What mass of aluminium is needed for the part? *(3)*
(e) Give two reasons why David always melts more than the minimum amount of aluminium. *(2)*

42 Carburettor control

This is a diagram of a carburettor. Use it to help you to answer the questions.

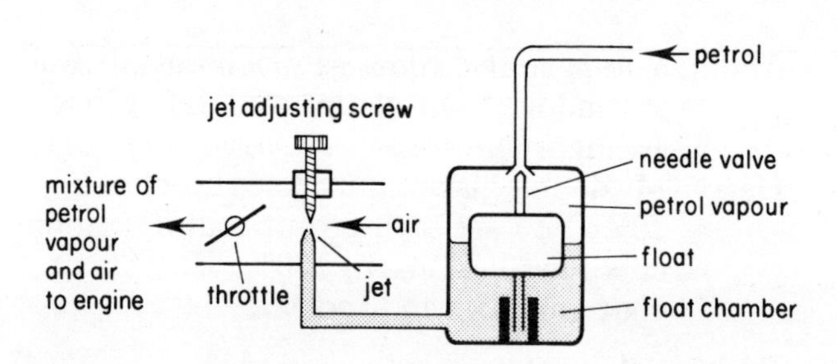

A carburettor controls the supply of petrol to a car engine.

(a) What does the needle valve do? *(1)*
(b) Why is the needle valve connected to a float? *(2)*
(c) Where is the mixture of petrol and air going? *(1)*
(d) What kind of problems are produced by a sticking needle valve? *(2)*
(e) What is the jet adjusting screw for? *(1)*
(f) In West Africa, many roads are not tarmacked so there is a lot of dust. This can get into the petrol tank and into the carburettor. Cars with dirt in the carburettor usually start easily in the morning but stop after travelling a few hundred metres. Explain this. *(4)*
(g) In a hot country, petrol evaporates easily. The pressure of petrol vapour in the float chamber is higher than in Britain.
    (i) How would this affect the performance of a car shipped out from Britain to Liberia? *(3)*
    (ii) How should the carburettor be adjusted to correct this? *(2)*

43 Costing weedkillers

Food production has increased in the last 40 years, partly due to crop protection by chemical sprays. Many of these sprays, both pesticides and herbicides, are made from fuels. They are by-products of oil refining. They contain energy themselves and are often flammable.

More energy still is needed to manufacture these crop protection chemicals. Here is a breakdown of the energy inputs needed to manufacture five farm chemicals used to protect grain crops. The energy measures are so large that they are in giga joules per tonne or GJ/t (giga = 1000 million).

Chemical name	*Energy input (GJ/t) needed to make chemical*						
	naptha	oil	gas	coke	electricity	steam	total
Paraquat	76.1	4.0	68.4	–	141.6	169.3	460
methyl parathion	37.0	2.0	24.0	6.0	73.0	18.0	160
Carbaryl	11.0	1.0	48.0	26.0	54.0	13.0	153
2-4-D	39.0	9.0		–	23.0	16.0	87
Captan	38.0	–	14	–	52.0	11.0	115

(Figures from the *Biologist* (1979, Vol 26, No. (3))

(a) Which chemical is likely to be the most expensive? *(1)*

(b) The weedkiller Captan can be sprayed between the rows of crops to kill weeds. The tractor needed uses 170 litres of fuel for the whole crop. 2-4-D, another weedkiller, can also be used, but a different tractor is required. It uses 340 litres of fuel for the whole crop. 0.2 tonne (200 kg) of weedkiller will be needed. The energy in 1 litre of fuel is 40 MJ (M = mega = million). Which weedkiller is the best choice to save the most energy altogether? *(2)*

(c) Methyl parathion, Carbaryl and Captan all start off with nearly the same energy input from naptha. So why do their final energy inputs vary so much? *(2)*

*44

In some caravans there is a layer of insulating material between the outer and inner walls. Caravans are a bit cheaper if the insulating material has aluminium foil on one side only. The manufacturers sell these caravans in England and in Italy. Diagram 2 shows how the cheaper insulation is fitted.

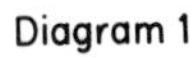
Diagram 1

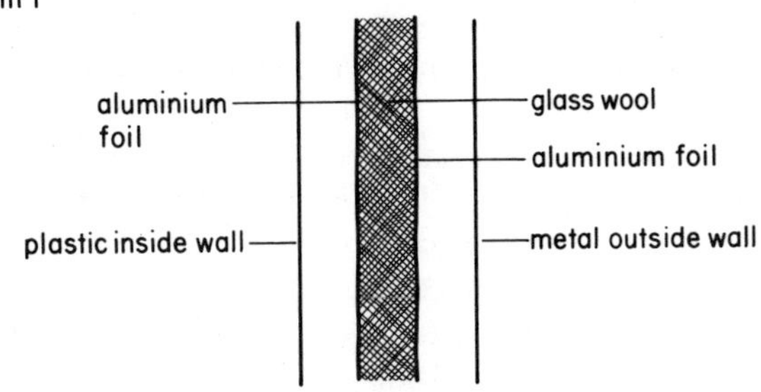

Diagram 2

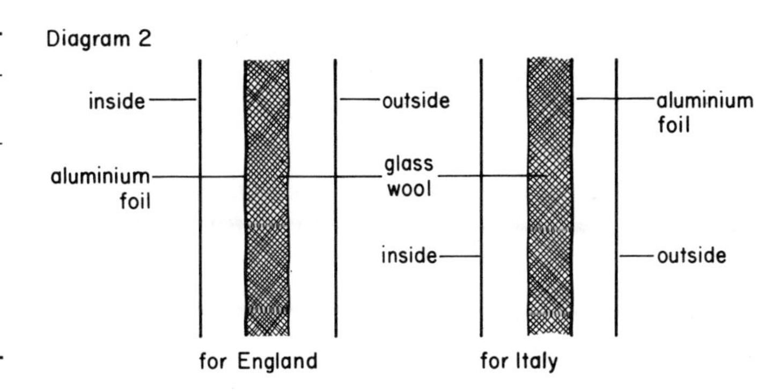

Are the two methods of using the cheaper insulation suitable for the climate in the countries where the caravans are sold? Explain your answer.

*45 Canning peas

This is a diagram of a system for sterilizing cans of peas. Use it to answer the questions below.

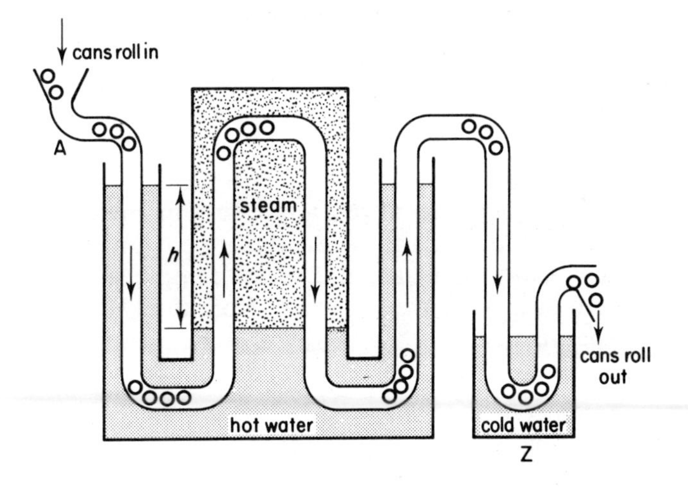

(a) The cans roll into the can sterilizer at A. They are warmed first in the water seal. Why should a closed can be warmed before heating in steam? *(2)*

(b) Where in the system will the cans be sterilized? What would the temperature be here? *(2)*

(c) What could be a danger if a can had a small hole in it? *(2)*

(d) What is the purpose of the hydrostatic head (this means difference of water levels) marked h? *(2)*

(e) The cooling pit at Z would need a control. What could that be? *(2)*

*46 Corroding copper

In moist air copper roofs turn green. The green substance is called verdigris. In clean air, the verdigris is basic copper(II) carbonate. In air contaminated with sulphur dioxide or hydrogen sulphide, the verdigris will contain some basic copper(II) sulphate. In air containing sea spray or hydrogen chloride, the verdigris will contain some basic copper(II) chloride. Once formed these coatings protect the metal from any further chemical action. Copper(II) carbonate gives off carbon dioxide when heated.

The dome of the Carnegie Library in Darwen, Lancashire, has been green for more than fifty years. Rashmi Patel is a science student who wants to test for pollution in the town. She scrapes 1 g of verdigris from the dome. She heats it strongly and collects all the gas given off. She adds 20 cm^3 of limewater. It goes milky.

She heats 1 g of pure basic copper(II) carbonate, collects all the gas given off and adds 20 cm^3 of limewater. It goes milky.

(a) The limewater from the experiment with verdigris is less milky than the limewater from pure carbonate. The student thinks this shows that there is pollution in the town. Explain her reasoning. *(2)*
(b) Rashmi must be very careful in the way she heats the samples if this is to be a fair experiment. Explain this. *(2)*
(c) Does the experiment really tell Rashmi anything about pollution in the town at the present time? *(2)*
(d) How could she make an accurate comparison of the milkiness of the limewater from the two experiments? *(3)*

47 Renovation

Brass is a mixture of copper (70%) and zinc (30%). It is a bright yellow metal. Brass drawer handles become tarnished. Red copper(I) oxide (Cu_2O) is first formed and then black copper(II) oxide (CuO).

Keith Holmes is an antique dealer. He needs to clean some handles which may have been on drawers for over a hundred years. He decides to take them off and drop them in vinegar. Vinegar is ethanoic acid. Copper does not react with ethanoic acid. Ethanoic acid reacts quickly with zinc and the oxides of copper.

zinc + ethanoic acid → zinc ethanoate + hydrogen

copper (I) oxide + ethanoic acid → copper (II) ethanoate + copper + water

copper (II) oxide + ethanoic acid → copper (II) ethanoate + water

Copper (II) ethanoate is soluble in water.

(a) (i) Will Mr Holmes' method for cleaning handles remove the tarnish? Explain your answer. *(2)*
(ii) Will Mr Holmes' method for cleaning handles affect the metal of the handles? Explain your answer. *(2)*

(b) Dilute ammonia solution reacts slowly with copper oxides and other salts to form a deep blue solution called Schweitzer's reagent. Neither zinc nor copper react with ammonia solution. Schweitzer's reagent dissolves cellulose (e.g. linen and rayon and other fabrics made from plant cells).
(i) How would you expect the use of dilute ammonia solution on tarnished brass handles to compare with the use of vinegar? *(2)*
(ii) Why would an old woollen sock be more satisfactory than cotton wool for polishing the brass with dilute ammonia solution? *(1)*

5 Synthesis

1 Synthesis

Phosphate ceramic is used by dentists for filling teeth. The impact strength of the material is tested by dropping balls made from it on to a hard surface. Adding fibres increases the impact strength of the ceramic.

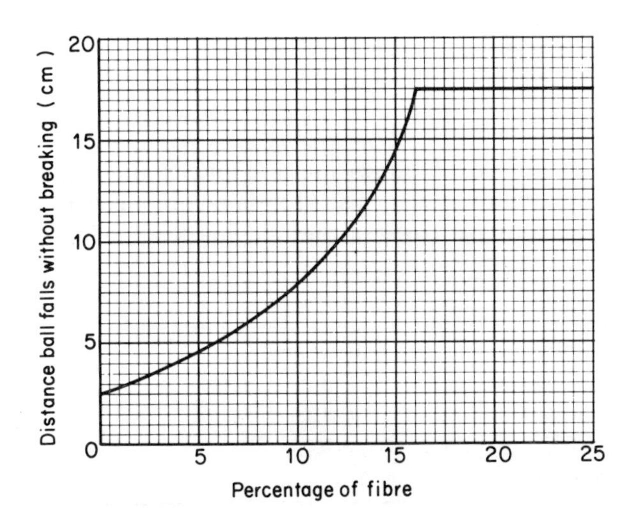

(a) How much fibre should be added to phosphate ceramic to make it as strong as possible? *(2)*
(b) Is the ball dropping experiment a reasonable test for a material which is to be used for filling teeth? Explain your reasoning. *(3)*

A

2 *U*-values

U-values are the heat losses through the fabric of a building, bricks, glass, metal and wood. They vary with the type of construction. Each construction material has a *U*-value. This is its ability to transfer heat (measured in watts per square metre) from the internal to the external air. The lower the *U*-value, the lower the loss of heat. Here are some *U*-values:

Single glazing with wood frames	4.3
Double glazing with wood frames	2.5
Single glazing with metal frames	5.6
Double glazing with metal frames	3.2
Tiled and felted roof, no insulation	1.5
Fully insulated roof	0.0

Calculating heat loss:

A = area of the part of the house to be measured (m^2)
H = rate of heat loss (watt)
T = temperature difference between external air and air in the room (°C)
U = known U-value

$$H = A \times T \times U$$

(a) Which type of glazing loses the most heat? *(1)*
(b) Is double glazing in a wooden framed window better than double glazing in a metal framed window? *(1)*
(c) How can you be sure that no heat is lost through the roof? *(1)*
(d) What would be missing from the roof with a *U*-value of 1.5? *(1)*
(e) What does a low *U*-value mean? *(1)*
(f) Calculate the rate of heat loss through a roof which is tiled and felted. It has an area of 100 m^2. The temperature in the roof is 40°C and outside the roof it is 30°C. *(3)*

*3 Yogurt making

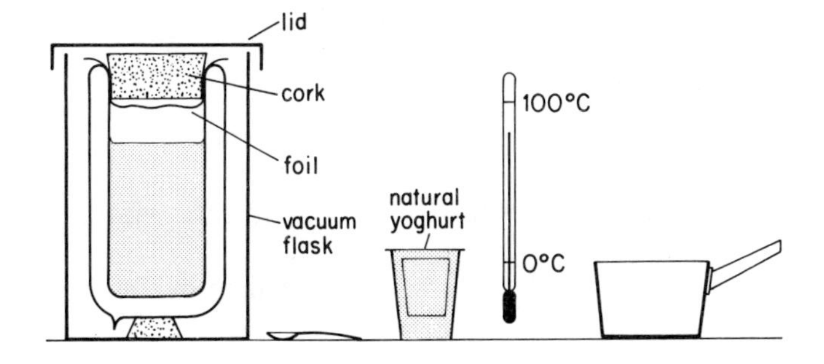

This question is about making yogurt at home. The equipment is shown in the diagram and the instructions are written out below. Look at the diagram and read the instructions, then answer the questions.

Instructions for making yogurt at home

(a) Collect together all of the equipment and sterilize it by washing it clean and soaking it in boiling water.
(b) Put the natural yogurt where you can find it.
(c) Warm ½ pint of milk to hand heat. Pour it into the vacuum flask and add two teaspoons of natural yogurt.
(d) Stir it in well.
(e) Fix the cork and foil. Leave at least three hours or overnight.
(f) You should now have a flask of natural yogurt.
 (i) Why should everything be sterilized? (*1*)
 (ii) How would you test the temperature of the milk? (*1*)
 (iii) What would this temperature be in °C? (*1*)
 (iv) Why do you need a vacuum flask? (*2*)
 (v) Read the instructions given for yogurt making at home and write them out as a flow chart. (*5*)

A *4 Seasonal change

This is a question about seasonal change. Here are two records of the same area. Use them to answer the questions below.

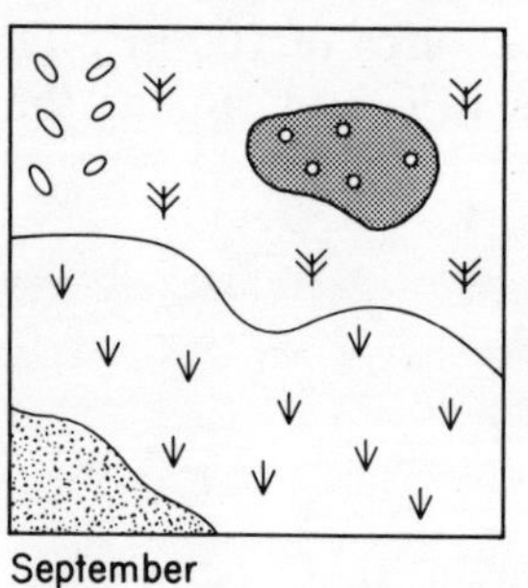

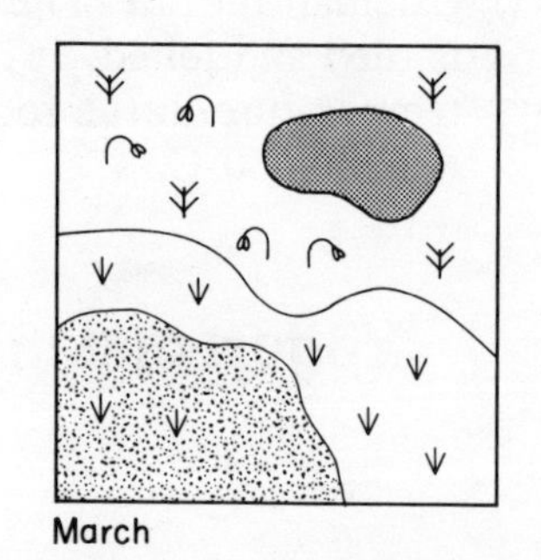

dead leaves | moss | grass | lichen | gravel | snowdrops | a piece of stone

(a) Which plant appears to have died over the winter? (*1*)
(b) Which plant shows a seasonal change in the area it covers? (*1*)
(c) Which plant shows no trace of its existence in September? (*1*)
(d) What other plant must be growing nearby? (*1*)
(e) Suggest a reason why these records are of a slope. (*1*)

*5 Blood test

Compare these two sets of data for blood tests and suggest possible reasons for the differences between them. The results are for humans.

Normal blood

Component	*Concentration*
red blood cells	5 million/mm^3
white blood cells	8000/mm^3
glucose	normal range 61–130 mg/100 cm^3
urea	normal range 14–40 mg/100 cm^3
temperature	33.5°C

John's blood test (early morning sample)

Component	*Concentration*
red blood cells	5 million/mm^3
white blood cells	12 000/mm^3
glucose	170 mg/100 cm^3
urea	18 mg/100 cm^3
temperature	38°C

6 Humus

Construct a flow diagram to show different stages of leaf decomposition as a fallen leaf becomes part of soil humus. This process is called humification. It involves two sets of changes. Leaf structure alters and molecular structures are broken down (degraded). Start with (a).

(a) Dead leaves fall to the ground and make leaf litter.
(b) The leaf cell enzymes begin to digest sugars, starch and simple proteins.
(c) Bacteria and fungi invade the leaf cells which are damaged. The bacteria and fungi digest cell contents.
(d) Leaves become broken up by woodlice and soil nematodes. Earthworms feed on the leaves.

The leaves become ground up in the worm's gizzard as it grinds soil particles with dead matter.

(e) Earthworm casts contain microscopic, semi-digested leaf fragments.
(f) Only strong leaf skeletons remain on the soil surface, or just below it.
(g) Other soil bacteria and fungi begin to attack the leaf skeleton, digesting any remaining cellulose.
(h) Earthworm activity mixes soil and leaf material.
(i) More soil bacteria and fungi attack the leaf stalk thickening (lignin) and any fatty or waxy molecules from the leaf cuticle.
(j) Finally, a sticky humus of organic molecules becomes linked to clay particles, so forming part of the soil humus.

*7 Oak tree food web

Use the information given below to construct a food web for an oak tree and its inhabitants. Begin at (a) but plan the web yourself. Remember that the arrows show the direction of energy flow.

Use ovals for links

Use boxes for ends of food chains

(a) Energy enters this food web as sunlight captured for the tree by leaves as they carry out photosynthesis.
(b) Squirrels eat acorns and green shoots.
(c) Birds eat caterpillars, greenfly, woodlice and occasionally centipedes.
(d) Soil bacteria decompose dead leaves.
(e) Lichens on the tree trunk use sunlight for photosynthesis and some minerals from rainwater on the bark.
(f) Caterpillars of the oak eggar moth feed on oak leaves.
(g) Parasitic insect grubs form spangle galls on oak leaves and oak 'apples' on twigs. They feed on plant tissue.
(h) Foxes eat squirrels and small birds.
(i) Fungi decompose dead leaves, branches and bark.
(j) Woodlice decompose bark.
(k) Centipedes prey on woodlice.
(l) Foxes have blood-sucking fleas.
(m) Rooks and magpies thieve birds' eggs.
(n) Mice eat acorns.
(o) Foxes and magpies feed on small mice.

8 Conifer planting

This table gives data about cultivated conifer trees.

	Growing conditions		
Trees	*Aspect*	*Heights* (m)	*Soil*
Sitka spruce	S, SE, SW	120–430	peat/heather ground
Norway spruce	S, SW	120–290	moister ground shallow peat
Douglas fir	S, SE	30–140	valley slopes and bracken with briars
Lodge Pole pine	W, NW	120–430	dry rocky land with poor soil

Work out a planting scheme for the heathery hillsides shown on the map. Copy the map and clearly mark the areas which could be used for planting each type of conifer.

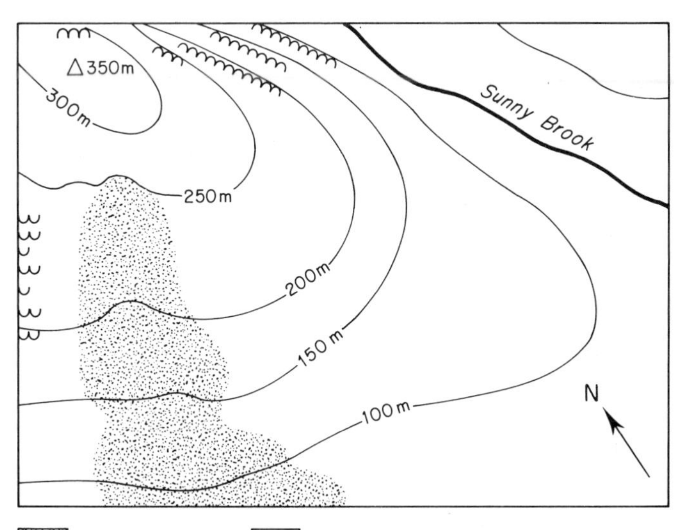

grass and rushes   rocks

9 Nuclear accident

Use the data below to compile

(i) a press report (7)
(ii) a flow diagram (3)

for the science column of a daily newspaper. Show how a major nuclear accident in one part of the world can affect farming economics and food production in different countries. Start with statement (a) but use the others as you wish.

(a) A nuclear reactor has suffered a major accident, releasing radioactive particles into the atmosphere.

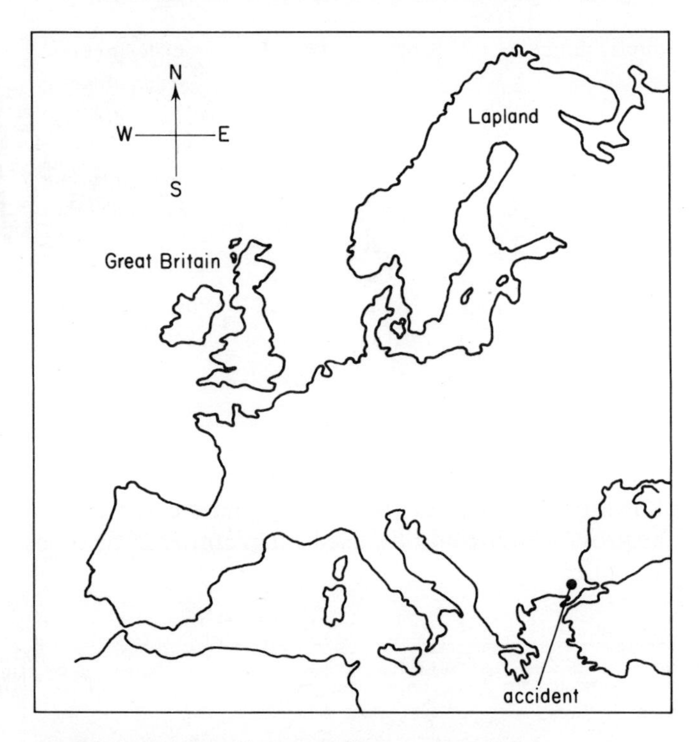

(b) Lichens absorb water like a sponge.
(c) Rainclouds condense around small particles.
(d) Weather patterns blow rainclouds from SE to N and NW.
(e) Reindeer feed almost exclusively on reindeer 'moss' – lichens.
(f) Animals such as sheep on upland pastures drink from mountain pools.
(g) Laplanders feed on reindeer meat.
(h) In Great Britain, hill farmers rely on lamb meat for part of their income.
(i) Laplanders must now import reindeer meat.
(j) Hill farmers must keep their lambs for longer until they have become less radioactive.
(k) Laplanders must destroy and burn all reindeer meat as its radiation count is well above safe limits.
(l) Radioactive elements may persist for about 30 years.

10 Salt sellers

Electrolysis of salt (sodium chloride) solution produces chlorine, sodium hydroxide and hydrogen.

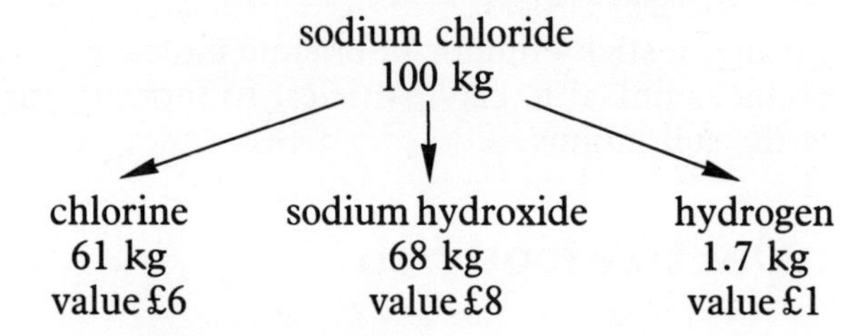

(a) (i) What is the total mass of the products? (2)
(ii) Why is the total mass of the products greater than the mass of salt? (1)

(b) A developing country has many salt lakes but no power stations. It buys electricity from a hydroelectric scheme in a neighbouring country. The electricity needed to electrolyse 100 kg of salt costs £7. The running of an electrolysis plant (wages etc.) to process 100 kg of salt costs £3.
(i) How much profit could be made from electrolysing 100 kg of salt if all products can be sold? (3)
(ii) A factory for electrolysing 50 million kilograms of salt per year could be built for a cost of £7½ million. How long would it take for this factory to pay for itself? (4)
(iii) There could be two major problems with this development. What are they? (2)

(c)

Product	*Uses*	*Other materials needed*
hydrogen	making margarine	vegetable oil (e.g. sunflower, soyabean, rapeseed, peanut, coconut)
	rocket fuel	
sodium hydroxide	making paper	softwood trees plus large quantities of water
	making soap	animal, vegetable or petroleum oil
chlorine	making PVC plastic	petroleum products
	making solvents	petroleum products

The country has many mountainous regions. Land between 2500 m and 4000 m started to be planted with quick growing larch and other softwood trees twenty-five years ago. The rainfall in the mountains is 90 cm per year. At high level there are many streams and rivers. But most rivers dry up before reaching the sea. Coconut palms are grown throughout the coastal plain. The country has no coal or oil and little industrial development.

(i) The country could develop three industries to use some of the products of salt electrolysis. What are they? Where should each of them be based? Explain your answers. (9)

(ii) Which of the products of salt electrolysis is likely to be a problem to get rid of? (1) Suggest how this problem could be solved. (1)

*11 Using zinc

Zinc is a commonly used metal. It is a fairly reactive metal, occurring between aluminium and iron in the reactivity series. The table gives some information about aluminium, zinc and iron.

Metal	*Melting point* (°C)	*Density* (kg/m³)
aluminium	660	2700
zinc	420	7140
iron	1540	7900

(a) The table below gives information about the industrial uses of zinc.

Use	*Percentage used*
galvanizing steel	40
manufacture of brass	20
sheet zinc	10
zinc alloy castings	25
other uses	5

(i) Construct a pie diagram for the data. (3)

(ii) In recent years the amount of brass manufactured has decreased. Why do you think this has occurred? (2)

(b) Many washing machine parts are made by pouring liquid zinc alloys into a mould and allowing the alloy to solidify. Why do you think manufacturers use zinc alloys in preference to aluminium or iron alloys? (3)

(c) A student noticed that many batteries use zinc as one of the electrodes. She also knew that two different metals can cause an electric current to flow. The student decided to find the voltage produced between zinc and other metals.

(i) Briefly describe how she would perform the experiment. (3)

(ii) The student's results are given below.

Metal	*Voltage* (volt)
aluminium	−0.9
copper	+1.1
iron	+0.3
lead	+0.6
magnesium	−1.6
silver	+1.6

(*Note* The negative voltage indicates that the current flows in the opposite direction)

Is there any similarity between these results and the reactivity series? (2)

(iii) The student noticed that in the experiment with copper, the zinc slowly disappeared, but it was the magnesium that disappeared in the experiment with that metal. Give a possible explanation for these observations. (3)

(d) (i) Suggest possible ways in which steel could be galvanized (that is, covered with zinc). (3)

(ii) Give possible reasons why zinc is used to protect steel from corrosion. (2)

*12 The monster

'The creature walked nearer to the children. It balanced on its hind legs, propping itself up with its thick tail to gaze at them. Its small front paws had sharp claws on the four fingers – more weapons to supplement its spiky teeth. There were fish scales around its mouth, contrasting with its leathery skin. Only on its back could the children glimpse armour plating glistening in the moonlight. No one had seen it out of the water before. They stared, transfixed by amazement as it removed three eggs from a pouch and buried them on the shore. Then it returned to the lake. As it swam away silently, the water surface was broken only by the wake from its snorkel tube and humpy back.'

(a) Draw a picture to show what this animal must be like. *(10)*
(b) Classify this creature. Give reasons. *(3)*
(c) Describe the monster's feeding habits. How are they linked to its habitat? *(3)*
(d) Could there be more than one monster? Explain your reasoning. *(1)*
(e) Write out a timetable of its probable daily routine. *(2)*
(f) Draw the animal swimming away from the children. *(2)*

*13 Growing seeds

Urma and Dawn tested three packets of seeds, A, B and C. They wanted to find out the best conditions to make the seeds grow. They knew already that the seeds should have air and water. They were investigating other factors, starting with temperature.

They tested three dishes, 1, 2, and 3, of seeds from each packet. Here are their results showing percentage germination (seeds growing out of 100) after a week.

Packet	*Temperature*		
	Dish 1 20°C	Dish 2 2°C	Dish 3 40°C
A	No seeds grew 0%	No seeds grew 0%	No seeds grew 0%
B	60% grew	20% grew	80% grew
C	No seeds grew 0%	2% grew	No seeds grew 0%

They were disappointed by these results and they left the dishes on the window sill at 20°C, just to see if any more seeds would grow. After three days, they noticed that seeds from packet A were growing in all of the dishes, 1, 2, and 3. The next day, the seeds from packet C in dish 2 began to grow well. Nearly all of them grew roots.

(a) How would you explain these results? *(3)*
(b) Design experiments to test your explanation. *(4)*
(c) Some seeds, such as willow-herb, need to have been exposed to light in order to germinate. Others, such as onion, will not germinate whilst they are in the light. Design an experiment to find out about how light affects particular seeds. *(3)*
(d) The behaviour of seeds C and those of willow-herb and onion may stop them from germinating at times or in places where they cannot grow. Explain this. *(3)*

*14 Fertilizers

Three essential elements for plant growth are nitrogen, phosphorus and potassium. The growth of plants is improved by artificial fertilizers which contain these elements in suitable compounds.

(a) Calculate the percentage of:
  (i) nitrogen in ammonium nitrate (NH_4NO_3),
  (ii) potassium in potassium sulphate (K_2SO_4).
  ($H = 1; N = 14; O = 16; S = 32; K = 39$) *(4)*
(b) Industrially, ammonium nitrate is manufactured by reacting ammonia and nitric acid. Describe briefly, how you would make a sample of pure ammonium nitrate in the laboratory, using ammonia solution and dilute nitric acid. *(4)*
(c) Streams and rivers are often polluted by nitrate ions. This may be due to excessive use of artificial fertilizers by farmers. A sample of water is taken from a stream on the first day of each month for a year. The amount of nitrate ions it contains is found.

Month	Jan	Feb	Mar	Apr	May	Jun	July	Aug	Sept	Oct	Nov	Dec
Nitrate ion ($\mu g/cm^3$)	5	4	15	16	13	12	12	12	15	35	10	8

  (i) Plot a histogram for these results. *(3)*
  (ii) What is the pattern? *(1)*
  (iii) Give possible explanations for the variations. *(3)*
  (iv) Suggest possible ways of reducing nitrate ion pollution. *(2)*
(d) In recent years it has become fashionable to grow foods organically. This means without the use of artificial fertilizers.
  (i) Why is it still essential for some form of fertilizer to be used? *(1)*
  (ii) What are the main sources of nitrogen fertilizer for organic farming? *(2)*
  (iii) Organic farmers add bone meal (crushed animal bones) to the soil. Why do they do this? *(1)*
  (iv) Give possible advantages and disadvantages of growing vegetables organically. Briefly describe how you might scientifically verify these. *(7)*
(e) For several months of the year there is frequent lightning around Lake Victoria in East Africa. It does not necessarily rain when this occurs. Lightning produces nitrogen oxides which dissolve in water to form nitric acid. Is this a good thing? Explain your answer. *(4)*

*15 Electromagnetic radiation

Diagram 1 gives information about the wavelengths of part of the electromagnetic spectrum.

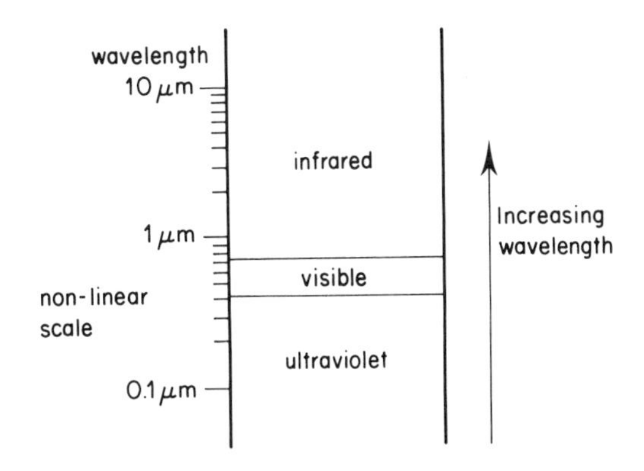

(a) (i) What is the longest wavelength of red light? *(1)*

(ii) What is the shortest wavelength of violet light? *(1)*

(iii) What is the wavelength of light in the middle of the visible spectrum? *(1)*

(b) Some telescopes which are used for studying stars have a mirror to concentrate the radiation (diagram 2). Graph 1 gives information about the amount of radiation reflected by different metal coatings.

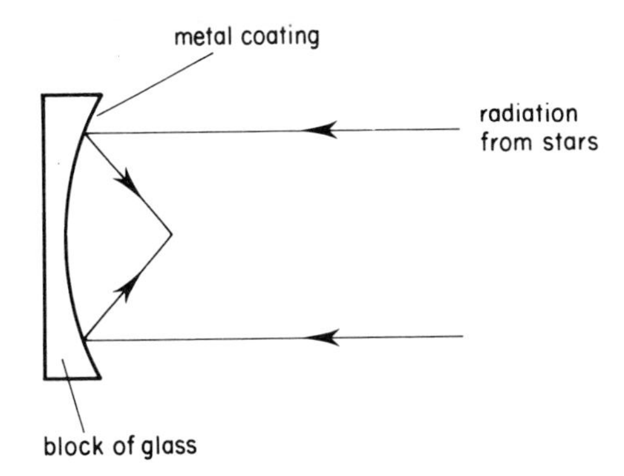

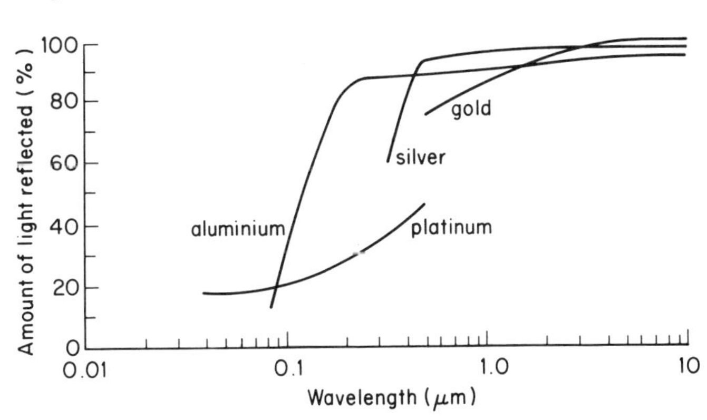

(i) Which material is the best reflector for short wave ultraviolet radiation? *(1)*

(ii) Which material is the best reflector for long wave infrared radiation? *(1)*

(iii) Which material is the best reflector for light in the middle of the visible spectrum? *(1)*

(iv) Which is the best material for a telescope which will be used for all radiation between the 0.1 μm and 10 μm wavelengths? *(1)*

(c) 'Below about 0.1 μm aluminium becomes transparent.' What could this statement mean? *(2)*

(d) In air, aluminium becomes covered with a very thin layer of oxide. This protects the surface from any further damage. What pattern, apart from their reflection properties, links the metals aluminium, platinum, silver and gold? *(1)*

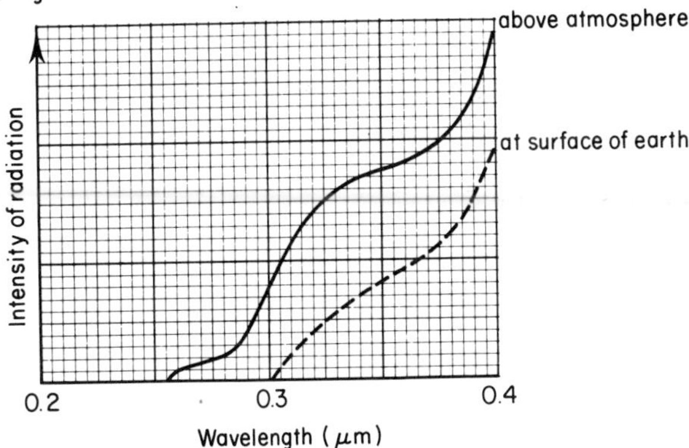

(e) Graph 2 shows how the intensity of the sun's radiation is affected by the earth's atmosphere.

(i) Describe the effect of the earth's atmosphere on the sun's radiation. *(3)*

(ii) Where should telescopes be built to study stars which emit most of their radiation at a wavelength of 0.33 μm? *(2)*

(f) Graph 3 shows which wavelengths give the best suntan.

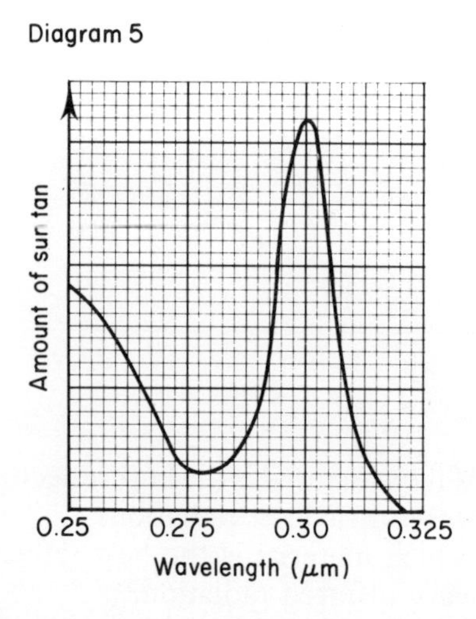

(i) Skiers can get sunburned inside their noses. Explain why. *(2)*
(ii) On the top of the Himalayan mountains (height 8000 m) where the temperature is −37°C, climbers can get very badly sunburned. Give two reasons for this. *(2)*
(iii) At a height of 8000 m, air pressure is only one-third of the air pressure at sea level. So Himalayan climbers are unlikely to have problems with sunburn inside the nose. Explain this. *(2)*

*16 Fly catcher

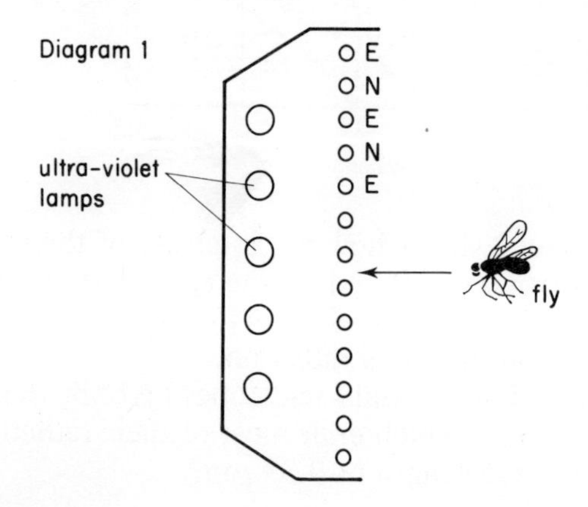

Diagram 1 shows a device for catching and killing flies in food shops. The rods marked E are at earth voltage. The rods marked N are at a large negative voltage (−1000 V). Insects fly towards the device when it is switched on. As they fly through the gap between E and N rods, a spark passes from one rod to the other. All insects flying between the rods are electrocuted and burned. The sparks convert normal oxygen molecules (O_2) to ozone (O_3). This produces a faint smell.

(a) Insects are good conductors of electricity. Air is a poor conductor of electricity. A 0.4 cm insect reduces the 1.0 cm air space between the rods to only 0.6 cm (diagram 2).

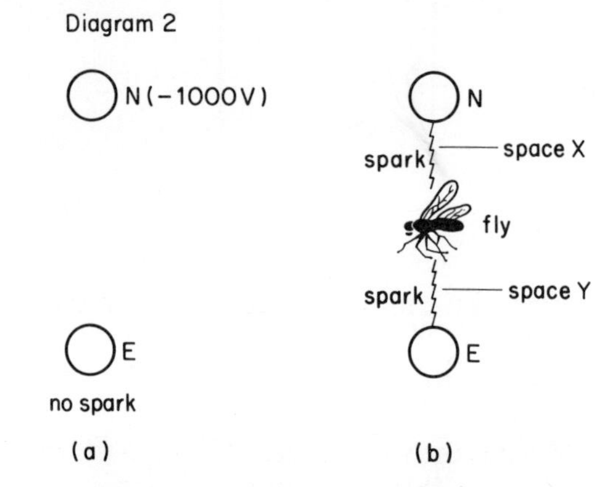

(i) What must be in the air space at X and Y for a current to be conducted? *(2)*
(ii) When the electric field between the rods is strong enough, the air 'breaks down' and a big current flows. The graph shows how current depends on the electric field (voltage per centimetre) when the device is working with the ultraviolet lamp on. Explain why breakdown can happen with an insect between the rods but not with just air between them. *(3)*

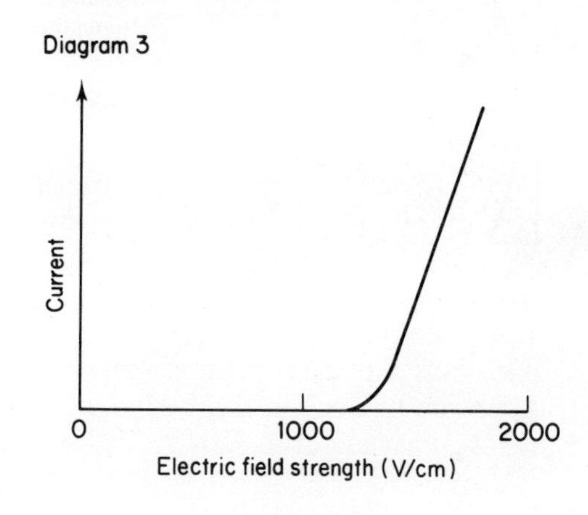

(b) The N rods can be at -1000 V with the UV lamp switched off.
(i) The number of insects flying towards the device falls from 50 per hour to 5 per hour. Explain this. *(2)*
(ii) Now, a few small insects can fly between the rods without being electrocuted. What does this tell us? *(3)*

(c) Read this passage. Then answer the questions which follow it.

Suntan is caused by ultraviolet radiation. Suntan can build up vitamin D in the body. Rickets is a disease caused by a shortage of vitamin D. Rickets normally affects children. It causes calcium deficiency and their bones do not harden properly.

Some fluorescent materials can absorb ultraviolet radiation and re-emit the energy they have absorbed as visible light. Some detergents contain fluorescent materials. The fluorescent materials remain in cotton clothes after washing. This helps to make the cotton look 'whiter than white'. The fluorescent materials do not remain in nylon clothes.

Ultraviolet radiation can kill germs. A wavelength of 0.29 μm is best. Air in hospital operating theatres is sterilized in this way. Ozone is a strong oxidizing agent. Like ultraviolet radiation, it can kill germs. It is used in Paris and Philadelphia for purifying water. A few parts per million in the air in cold stores can help preserve meats.

(i) Betty used to work in a post office. Her face used to be very pale. After taking a job in a shop with one of these insect catchers, her face looked 'healthier'. It was almost as brown in winter as after her summer holiday. Explain this. *(1)*
(ii) Where there is one of these insect catchers, ladies need to take care what they wear. It is not a good idea to wear a thin white nylon blouse over white cotton underwear. There is a particular problem on a dark day. Explain this. *(4)*
(iii) Streptococci bacteria cause sore throats. The germs are spread by coughs and sneezes. Community Health Inspectors find that there are fewer streptococci bacteria where these insect catchers are used. Explain this. *(3)*
(iv) John is sixteen. He has acne on his face, back and chest. For a few years after puberty, the hormone balance in the body can produce a lot of greasy material. When oil glands in the skin get blocked they can be infected by staphylococci bacteria. As the spots burst, the infection spreads to cause other spots. After working under an insect catcher for a few weeks, there are fewer spots on John's face. But his back and chest are as bad as ever. Explain this. *(3)*

*17 Rain lines

Hot sodium produces yellow light. It emits two wavelengths (0.5890 μm and 0.5896 μm). This yellow light can be seen when salt (sodium chloride) is scattered in a flame. White light from an electric lamp normally produces a complete spectrum of colours. If this white light passes through cool sodium vapour yellow light (of wavelengths 0.5890 μm and 0.5896 μm) is absorbed. Two dark lines are produced in the spectrum.

(a) Draw a diagram to show how a prism could be used to form a spectrum from white light. Label the red and blue light produced. *(4)*
(b) Diagram 1 shows a star. Light from a star can be used to produce a spectrum. There are several dark lines in the spectrum. They are called Fraunhofer lines (diagram 2). The wavelengths of these dark lines are given in the table.

Diagram 1

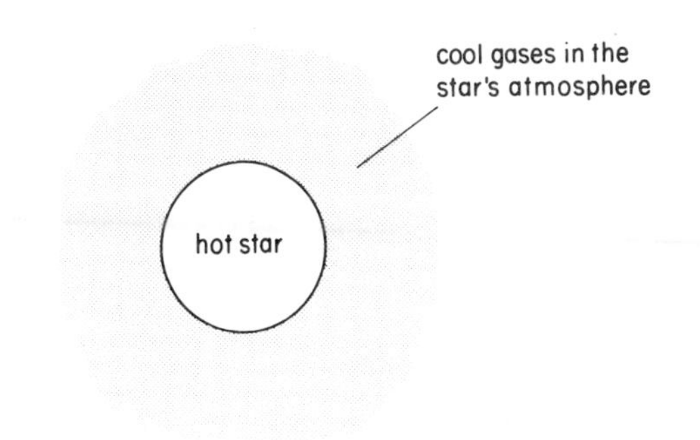

Diagram 2

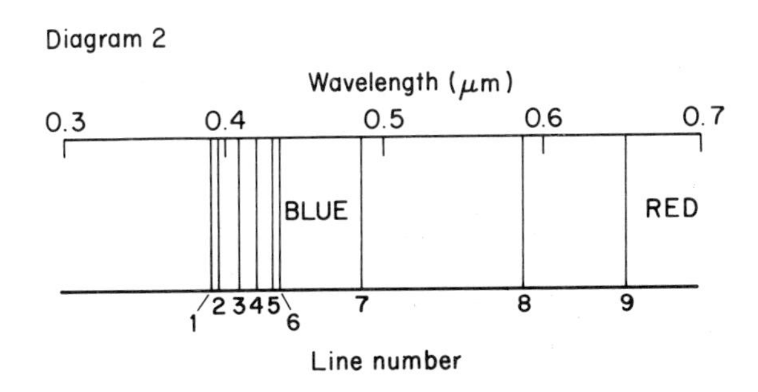

Line number	*Wavelength* (μm)
1	0.3934
2	0.3968
3	0.4102
4	0.4227
5	0.4308
6	0.4340
7	0.4861
8	0.5876
9	0.6563

Here are the wavelengths produced by some elements when they are hot:

Element	*Wavelength* (μm)
calcium	0.3934 0.3968 0.4227 0.4308
helium	0.5876
hydrogen	0.4102 0.4340 0.4861 0.6563
magnesium	0.5167 0.5173 0.5184
oxygen	0.6276–0.6287* 0.6867–0.6884* 0.7594–0.7621*

* Band

(i) Which elements are in the atmosphere of the star? *(3)*

(ii) Which of these elements would you expect to find near the star rather than at the outer edge of the atmosphere? Explain your reasoning. *(2)*

(iii) If sunlight reflected by the moon is examined from a spaceship, dark bands are seen at the oxygen wavelengths. But the moon has no atmosphere. What do the dark bands tell us? *(1)*

(c) Some dark lines in the sun's spectrum become much darker when the sun is low in the sky. They are very dark at sunrise and sunset. These lines are darkest on humid days. But they are much less dark when the spectrum is examined at a height of 3000 m. The same lines are seen when a white light is observed through 35 m of steam. These lines are called Telluric lines.

(i) What causes Telluric lines? *(2)*

(ii) Why are they weak at a height of 3000 m? *(2)*

(iii) Why are they very dark at sunrise and sunset? *(2)*

(iv) Explain how Telluric lines could be used to forecast the weather. Describe exactly what should be done to make forecasts as reliable as possible. *(4)*

(v) Could Telluric lines be used to forecast the weather during the night? Explain your answer. *(2)*

*18 Damp walls

Sometimes the lower part of housewalls is damp on the inside. This is due to rising damp. Brick, stone and plaster can soak up moisture from the ground. They act like blotting paper. A waterproof layer called a damp proof course (d.p.c.) is supposed to prevent this. But sometimes the damp course is damaged.

This soaking-up effect is called capillarity. It happens when a narrow tube is dipped in water (diagram 1). The graph shows the height reached by water in tubes of different diameters.

Diagram 1

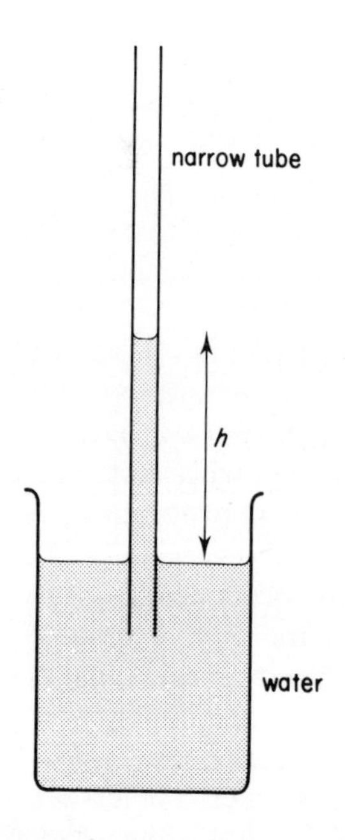

(a) (i) Rising damp comes 75 cm up a wall. What is the average size of the space between particles in the building material? *(1)*

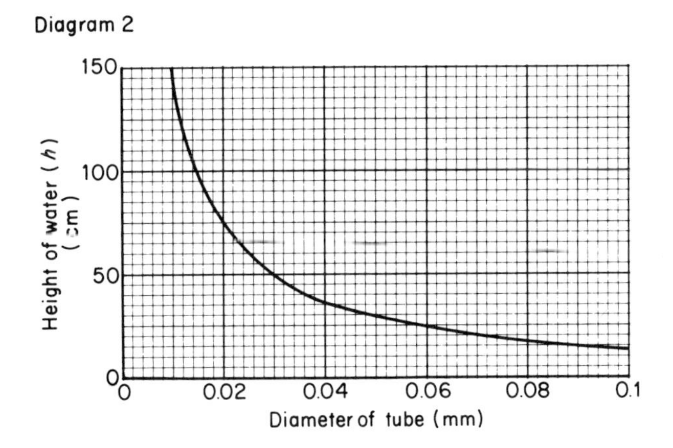

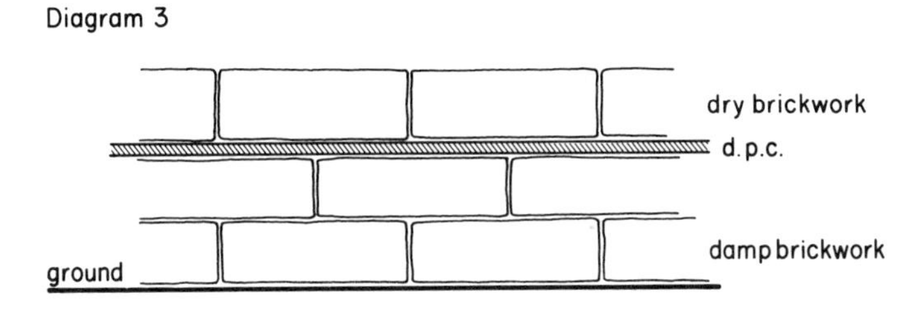

(ii) In older houses, the d.p.c. is a layer of slate. But slate must be different from other kinds of stone if it can stop rising damp. In what way must it be different? *(1)*

(iii) Name two other solid materials which could be used as a d.p.c. *(2)*

(iv) Some older buildings do not have a normal d.p.c. It is possible to stop dampness rising by drilling holes in the wall and squirting a liquid into the stone or brick. The liquid later goes solid. How must this treatment alter the stone or brick? *(1)*

(b) A number of different salts can be found in building materials. One of these is calcium chloride ($CaCl_2$). Calcium chloride is hygroscopic. This means that it attracts water. It forms a hexahydrate, $CaCl_2.6H_2O$.

(i)

Element	*Relative atomic mass*
calcium	40
chlorine	35.5
hydrogen	1
oxygen	16

Use this information to calculate how much water there is in 1 g of $CaCl_2.6H_2O$ *(4)*

(ii) Diagram 4 shows what happens if blotting paper is dipped in a solution of calcium chloride. Why does the water rise 10 cm with this blotting paper? *(2)*

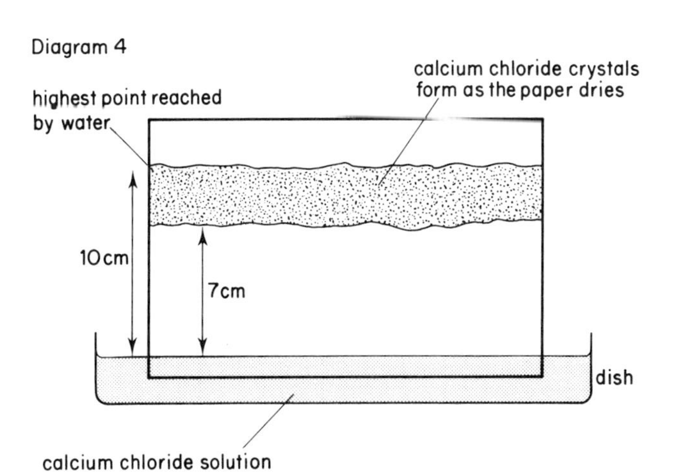

No calcium chloride is deposited in the first 7 cm above the level of liquid in the dish. It is only deposited between 7 cm and 10 cm above the level of liquid in the dish. Where the calcium chloride is deposited depends on two things. What are they? *(2)*

(iii) Diagram 5 shows where crystals of calcium chloride are deposited on some brickwork inside a house. What does the pattern on the wall tell us? *(4)*

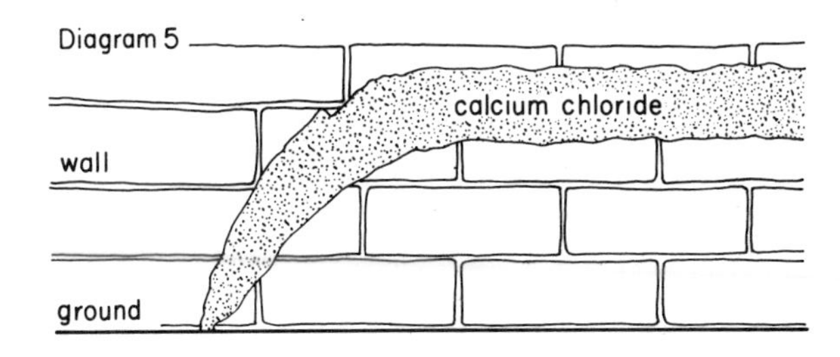

(iv) Remember that calcium chloride is hygroscopic, that is, it attracts water. Go back to (a)(i). Do you think your answer would be correct for a wall which had calcium chloride deposited in it? Explain your reasoning. *(3)*

(c) Glass is hygroscopic. Its surface attracts water. When glass is ground up, the glass powder can absorb a great deal of water from the air. Glass is made from silica. Most sand is silica. Like glass powder, sand absorbs water from the air. Explain why this can cause a problem in houses built of sandstone. *(3)*

*19 More damp walls

(a) After a cold night, bedroom windows are often covered with droplets of water. This is called condensation. Condensation on windows is unusual in summer. Condensation can be a particular problem in rooms containing a lot of plants.
  (i) Why do plants make condensation worse? (*1*)
  (ii) Damp walls can be caused by water vapour in a room condensing on a wall. Which line in the table below gives the conditions under which condensation on walls is likely to happen? (*1*)

	Temperature of wall	*Temperature of air in room*	*Humidity of air in room*
A	warm	cold	dry
B	cold	warm	dry
C	cold	warm	moist
D	warm	cold	moist

(b) There are spores of many types of fungus floating about in the air. The spores need moisture to germinate. The mould needs moisture to survive and grow. Some wallpaper paste contains chemicals which kill moulds. The chemicals are fungicides. In which rooms of a house is it important to use paste containing fungicides? (*3*)

(c) House surveyors have a dampmeter which shows whether or not there is dampness in a wall. The dampmeter contains an ammeter. The maximum current which can be shown on the ammeter is 0.1 A. When the metal pins are pushed into a damp wall, the device shows a current.

Diagram 1

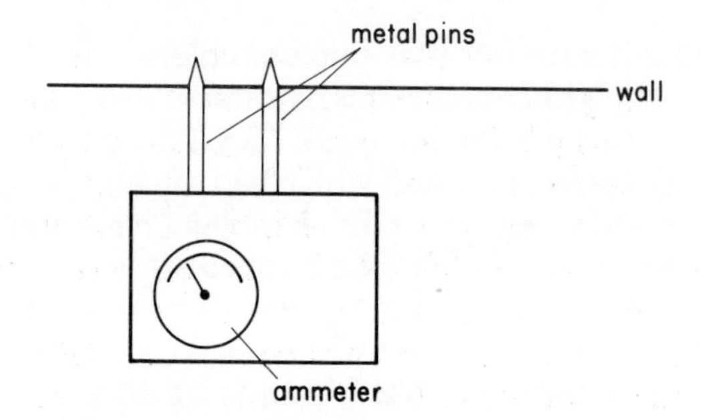

  (i) What else must this dampmeter contain besides an ammeter? Draw a circuit diagram. Calculate the specifications (for example, resistance, e.m.f., wattage) of the components used. (*5*)
  (ii) Do you think the reading on the dampmeter depends on whether there are salts in the plaster? Explain your reasoning. (*3*)
  (iii) After many years of use, one pin of these dampmeters is thinner than the other. It is always the same pin. Explain this. (*2*)

(d) Some people put a layer of aluminium foil under the wallpaper. This stops dampness in the wall damaging the wallpaper. The wall side of the aluminium foil is covered with tar.

Diagram 2

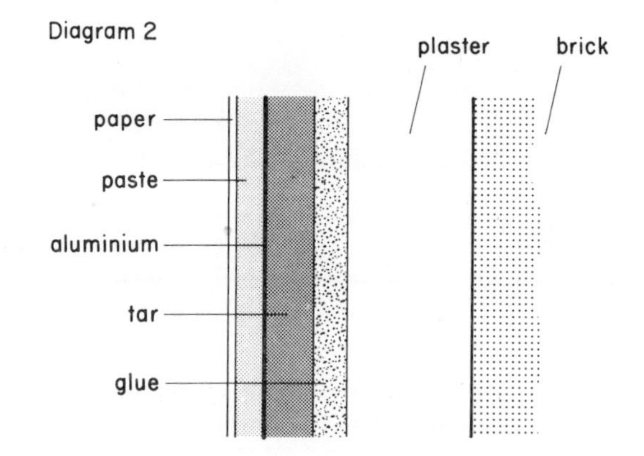

  (i) Explain two ways in which the tar is useful. (*2*)
  (ii) What reading would you expect on the dampmeter described in (c) if it is pressed into a wall covered with this material on a day when the wall is dry? (*1*)

(e) In 1809, F. Reuss discovered electro-osmosis. Diagram 3 shows what he did. Explain how electro-osmosis could be used to solve the problem of rising damp in a wall. (*4*)

Diagram 3

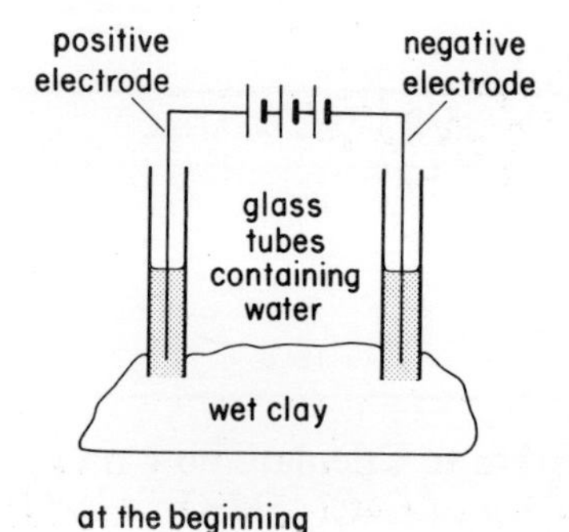

at the beginning

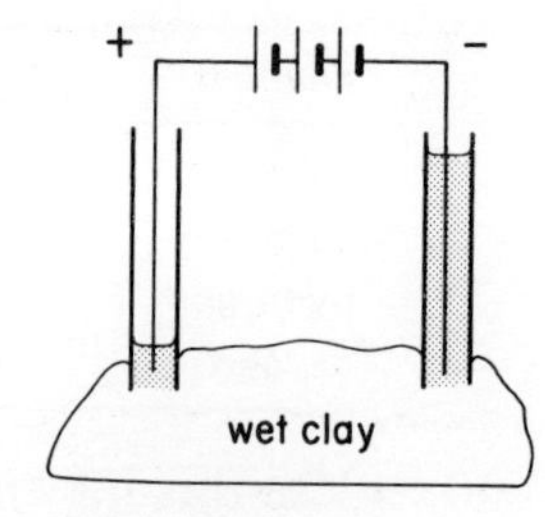

after a few hours

20 Wadi Halfa

Mr El Darwish lives in Wadi Halfa in the Sudan. In the middle of the day the temperature is usually about 42°C. It only rains about once every ten years.

Diagram 1

(a) He wears a white kanzu. This is made of very thin cotton. It is loose fitting and hangs from his shoulders down to his feet like a very long shirt. On his head he wears a white turban. His wife wears a 6 m length of black cotton wrapped round the body and over the head. How suitable are these clothes for the climate? Evaluate the clothing in terms of heat transfer. *(6)*

(b) Mr El Darwish has a large pot outside his door. It is unglazed so that water can soak through to the outside. His servant brings water from the river Nile to fill the pot each day. The graph shows how the temperature of the water changes. The water is refreshing at 25°C or below.

Diagram 2

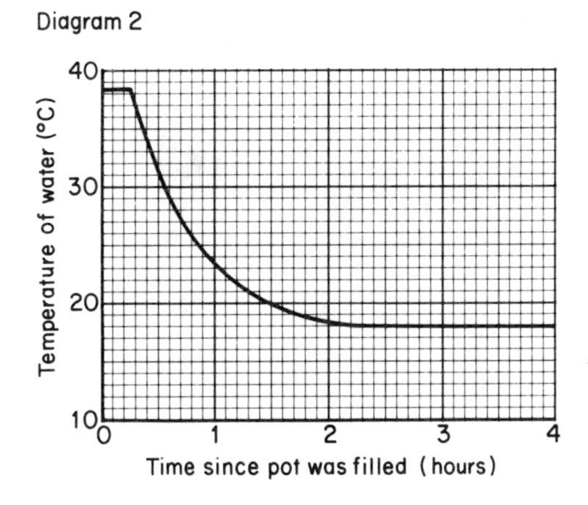

(i) How long does it take for the water to become pleasant to drink? *(1)*
(ii) What causes the temperature of the water to change? *(2)*
(iii) Why does it take 15 min before the temperature begins to change? *(1)*
(iv) Eventually the temperature stops changing. Why is this? *(2)*

(c) There is a metal tank on the roof of the house. It provides water for a shower. The servant fills the tank with 30 litres of water each morning. The graph shows how the temperature of the water changes. The water is too hot for a shower if it is above 50°C.

Diagram 3

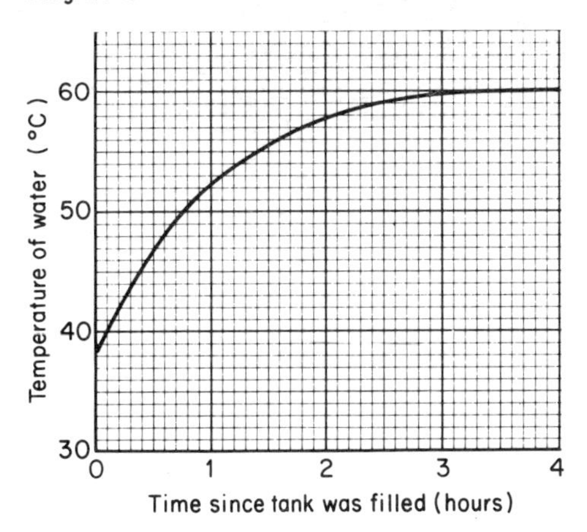

(i) How much time does Mr El Darwish's family have to use the shower before the water becomes too hot? *(1)*
(ii) What causes the temperature of the water to change? *(1)*
(iii) Eventually the temperature stops changing. Why is this? *(2)*

(d) (i) Mr El Darwish drinks Nile water from his pot all the time. It is full of germs but he does not often become ill. Why is this? *(2)*
(ii) A foreign visitor, Mr James, is offered a glass of Nile water but says that he would be pleased to take some tea instead. Why is this a good idea? *(3)*
(iii) Another foreign visitor, Mr Choudhary, says that he would be honoured to drink a glass of Nile water but would Mr El Darwish mind if he took some medicine with it. He drops two pills in the water. After a minute or two Mr Choudhary drinks the water. The pills of course make the water safe to drink. What might they contain? *(2)*

21 Conception

An egg is released from one of a woman's two ovaries about once every 28 days. This is called ovulation. The egg passes down an egg tube (Fallopian tube) to the womb (uterus). The egg may take a few days to pass down the egg tube or it may take only a few hours.

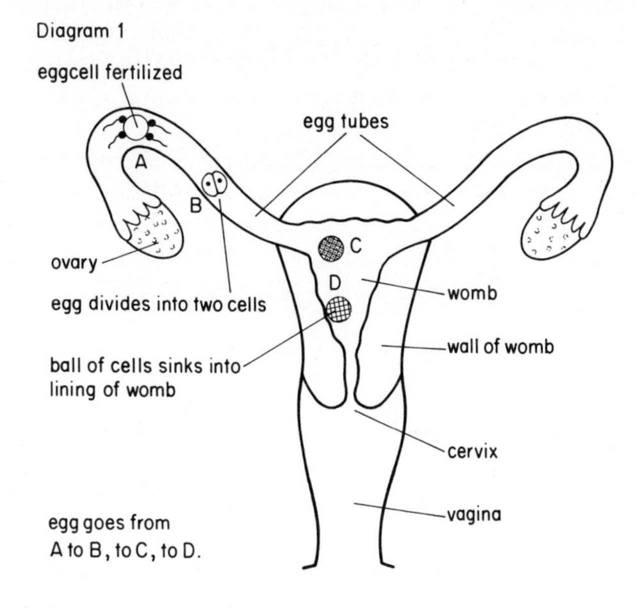

Diagram 2

menstruation
egg released but not fertilized
menstruation
egg released and fertilized
PREGNANT
Womb lining
Oestrogen
Progesterone
sore throat
Temperature (°C)
38
37.5
37
36.5
36
0
14
28
42
56
70
Number of days

For a baby to be conceived, sperm from a man must reach the egg while it is passing down the egg tube. This is called fertilization.

When an ovary releases an egg, there is a change in the hormone balance in the woman's body. There is also a change in her body temperature. Diagram 2 shows the changes which usually take place. The wall of the womb thickens as it prepares to receive a fertilized egg. If the egg is not fertilized, the womb lining comes away. This causes bleeding from the vagina (menstruation).

(a) How many eggs might a woman release between puberty at the age of twelve and the menopause at the age of forty-eight? *(3)*

(b) (i) What happens to the amount of progesterone in the body just after ovulation? *(1)*
(ii) When does the amount of oestrogen increase? *(1)*

(c) The diagram shows when an egg is released by the ovary. An arrow shows that the egg goes to the wall of the womb.

(i) How long does it take for the egg to reach the wall of the womb? *(1)*
(ii) Sperm can remain alive inside a woman for about 2 days. At what times during the woman's monthly cycle is intercourse most unlikely to result in a pregnancy? *(2)*

(d) (i) Describe the pattern of temperature changes during a woman's monthly cycle. *(3)*
(ii) What kind of things can upset this pattern? *(1)*

(e) Sometimes a man and woman try to have a baby for several years without the woman becoming pregnant.
(i) This may be because they do not often have intercourse at the best time during a woman's monthly cycle. This can be a serious problem if eggs remain in the woman's egg tube for only a few hours. Explain how the use of a medical thermometer could help such people. *(2)*
(ii) The vagina is slightly acid. The mucus of the cervix is usually alkaline. Sperm have less chance of surviving in an acid environment than in an alkaline one. During intercourse, the man deposits about 4 cm^3 of semen in the woman's vagina. The semen is a fluid containing sperm. There is usually a pool of semen in the vagina after intercourse. For the woman to become pregnant, sperm must swim through the cervix, through the mucus inside the womb and up the egg tube.

Diagram 3

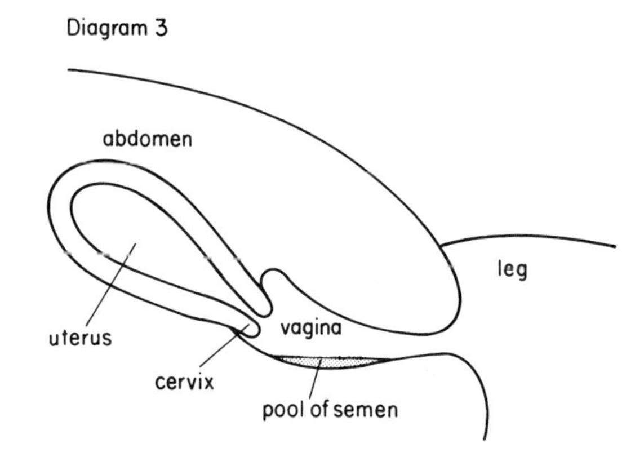

Many years ago, women were sometimes advised to rinse their vagina with sodium hydrogen carbonate solution before intercourse. It was thought that this would increase the chance of pregnancy. Explain whether or not this practice is likely to help solve the problem. *(3)*

The chance of sperm reaching the egg tube may depend on the position of the woman's body. Explain this. *(2)*

(iii) Sperm are produced in a man's testicles (testes). The temperature of the main part of the human body is 37°C. Sperm are best produced at a temperature several degrees cooler. The testicles hang down from the body in a sac of skin called the scrotum. There are muscles in the scrotum. When these muscles contract, the testicles are pulled up close to the body. When the muscles relax, the testicles hang away from the body. There are a lot of sweat glands in the scrotum. The evaporation of sweat affects the temperature of the testicles.

Diagram 4

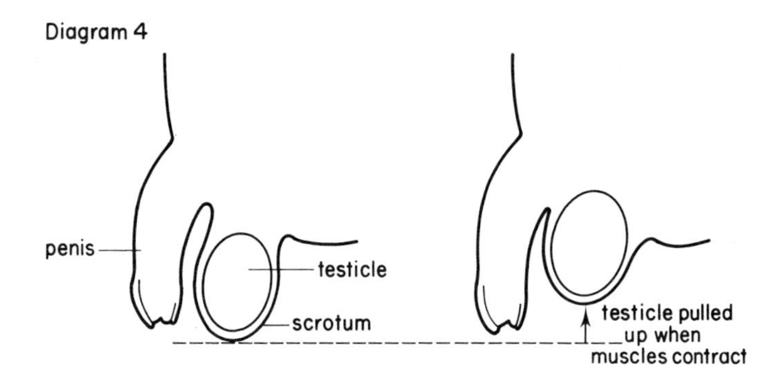

Explain how a man's body may use the muscles and sweat glands in the scrotum to keep his testicles at the best temperature for making sperm. *(3)*

If a man and woman have not been successful in producing a pregnancy, it may be because the man's testicles are not making sperm properly. This problem could be caused by the kind of clothing the man likes to wear. Explain how the problem may be solved. *(2)*

*22 Resourceful limestone

When limestone (calcium carbonate, $CaCO_3$) is heated, it decomposes into quicklime (CaO) and carbon dioxide (CO_2). This reaction takes place in a calcinating kiln.

$$CaCO_3 \rightarrow CaO + CO_2$$

Calcium carbonate reacts with dilute acid. For example:

$$CaCO_3 + 2HCl \rightarrow CaCl_2 + H_2O + CO_2$$

calcium carbonate + hydrochloric acid → calcium chloride + water + carbon dioxide

(a) Powdered limestone is dusted about in coalmines. This prevents the spread of flames if there is a fire. Explain this. *(2)*

(b) Finely powdered calcium carbonate is a white powder. It is used in some toothpastes.
   (i) Powdered calcium carbonate helps to prevent tooth decay and gum disease. Explain this. *(5)*
   (ii) What happens if the calcium carbonate is swallowed down to the stomach? *(3)*

(c) Quicklime is used in very bright arc lights. These used to be used in theatres and were called 'limelights'. Limelights produce several colours of the spectrum and these give the effect of white light. What are the three main colours if they mix to make white light? *(3)*

(d) Quicklime reacts with water to give slaked lime (calcium hydroxide, $Ca(OH)_2$.) This is a lively reaction and a lot of heat is produced. During this slaking process, the material expands to twice its volume.
   (i) The slaking of quicklime can be used for splitting stone. Give two reasons why this works well. *(2)*
   (ii) The human body is mostly water. Sometimes thousands of people die because of earthquakes or other disasters. Large pits are

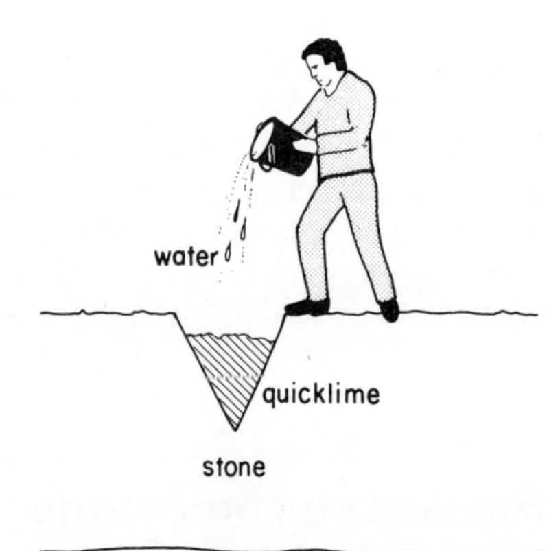

dug, the bodies are placed in the pit and covered with quicklime. Other bodies are then collected and added to those already in the pit. The lime decomposes the bodies. It also kills germs. Explain how the lime works. *(3)*

(e) Slaked lime absorbs carbon dioxide from the air to form calcium carbonate.

$$Ca(OH)_2 + CO_2 \rightarrow CaCO_3 + H_2O$$

Until the invention of modern cement, lime mortar was used in building. This is a mixture of slaked lime and sand. The mortar absorbs carbon dioxide from the air and water evaporates from it. Explain why the mortar goes as hard as stone. *(3)*

*23 Antifreeze

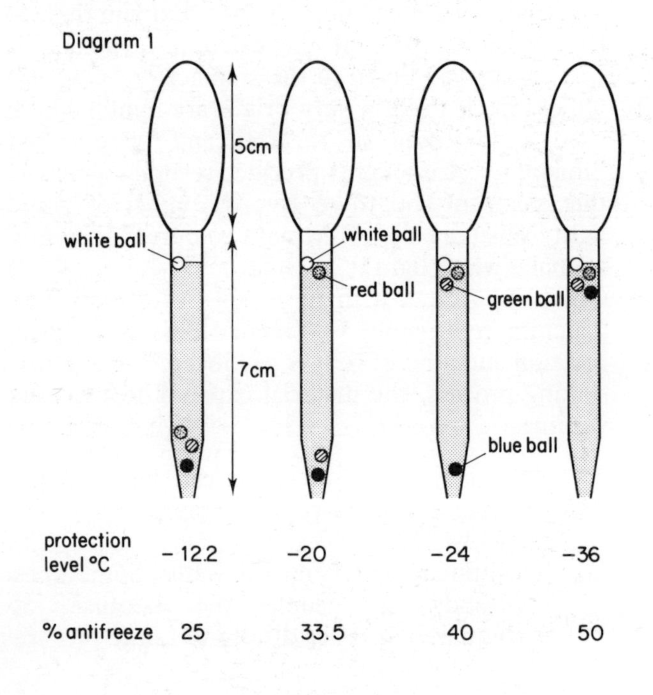

Ethylene glycol is used in car radiators as an 'antifreeze'. This protects the radiator in cold weather. An antifreeze tester consists of a syringe containing four balls. The balls are all the same size. Some of the instructions are shown in Diagram 1.

(a) The liquid in a car radiator may be 12 cm below the radiator cap. How could you get the liquid into the syringe? *(2)*

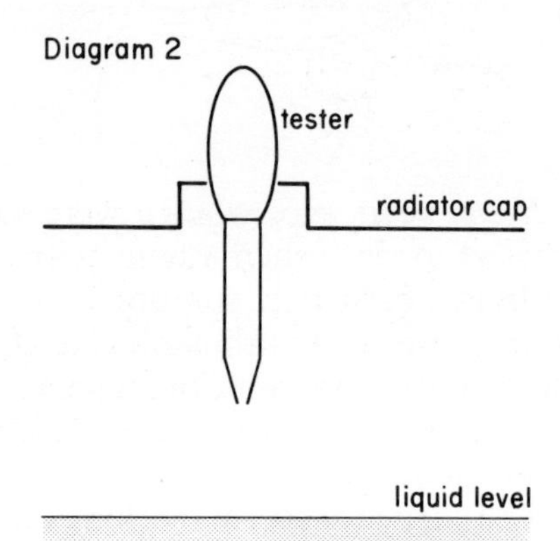

(b) (i) If there is no antifreeze in the radiator, it will freeze at 0°C. Draw a graph of protection level against percentage of antifreeze. The graph should have five points. *(5)*

(ii) One of the pieces of information in the instructions may be incorrect. Which is it? *(1)*

(iii) How much protection would be given by radiator water containing 20% ethylene glycol? *(1)*

(c) Which is the lightest ball? *(1)*

(d) Is ethylene glycol denser or less dense than water? Explain your reasoning. *(3)*

(e) How many balls would you expect to float in pure water? *(1)*

(f) The graph shows how the volume of water depends on temperature.

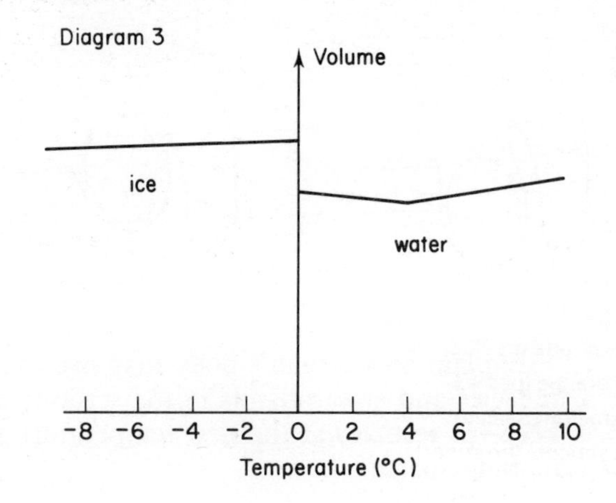

(i) If the radiator is filled with pure water, at what temperature will the level of water in the radiator be lowest? *(1)*

(ii) A radiator is filled with pure water, but not all of it is at the same temperature. The coldest is 1°C and the warmest is 3°C. Where would the 3°C be found? Explain your reasoning. *(3)*

(iii) If the radiator is filled with pure water, where will ice first start to form? *(1)*

(iv) What would happen to a radiator if the water inside it froze? *(1)*

24 Menu making

A member of your family has to follow a diet because of high blood sugar levels. The diet will control blood sugar in two ways. It cuts down the amount of carbohydrate eaten while high fibre fills up the person, preventing hunger pangs. High fibre regulates the flow of sugar into the blood from digestion.

Plan two daily menus: Menu 1 is an everyday menu, Menu 2 is for a celebration day.

Rules

(a) Energy count should be about 1500 kJ per meal.
(b) Daily fibre intake should be between 40–50 g.
(c) There must be one piece of fruit per day for vitamin C intake.
(d) There must be 110–170 g of protein per day.
(e) Frying doubles energy content.
(f) Microwaving, boiling and roasting do not add energy unless sugar or fat are added.
(g) Meats contain no fibre.

Food	*Energy* (kJ)	*Protein* (g)	*Fibre* (g)
40 g bran cereal + milk	670	5	12
100 g jacket potato	300	2	2.5
200 g baked beans (slimline)	450	10	12.5
100 g red beans	950	15	7.1
150 g tinned peas	700	6	11.2
50 g uncooked wholewheat pasta	1680	7	5.7
100 g apple (1 apple)	310	–	2.2
100 g pear (1 pear)	175	–	1.7
100 g banana (peeled)	755	1	3.4
100 g orange (peeled)	240	1	2.5
200 g melon (peeled)	200	2	1.4
100 g apricots (no sugar)	170	1	10.0
(continued)			
100 g blackberries	170	1	8.3
100 g boiled greens	65	2	4.3
100 g boiled green beans	65	2	3.9
100 g raw mushrooms	30	2	2.5
100 g cauliflower	65	2	2.0
50 g bean shoots	60	2	1.7
100 g canned tomatoes	75	1	1.0
2 slices granary bread	755	7	4.4
2 slices white bread	750	6	4.4
100 g cheese	1680	26	0
50 g bacon (streaky)	870	7	0
50 g beef (lean)	500	10	0
50 g lamb	700	8	0
50 g pork	670	8	0
50 g corned beef	450	13	0
50 g sausage (beef)	620	5	0
1 egg	305	6	0
50 g white fish	160	9	0
100 g whole milk	270	3	0
100 g skimmed milk	145	4	0

25 Other diets

In the Far East, the average person eats about 182.5 kg of rice per year. In Britain and America the figure is about 3 kg. Rice is mostly carbohydrate. It is a source of energy. It contains 15 kJ/g. The energy needed by a factory worker is 12 000 kJ/day. The energy needed by a child of six is 7500 kJ/day. An average person can survive satisfactorily with 10 000 kJ/day.

(a) (i) How much rice does the average person in the Far East eat each day? *(2)*

(ii) How much energy does he get from this? *(2)*

(iii) What percentage of the necessary daily intake of energy comes from rice? *(1)*

(iv) Extra energy could be taken in by eating other foods. Name three other foods which contain a lot of carbohydrate. *(3)*

(b) 27% of the people in the Far East are 'starving'. (They get less than 9200 kJ/day). Copy and complete the graph to show the likely daily intake of the 73% of the people in the Far East who get more than 9200 kJ/day. *(3)*

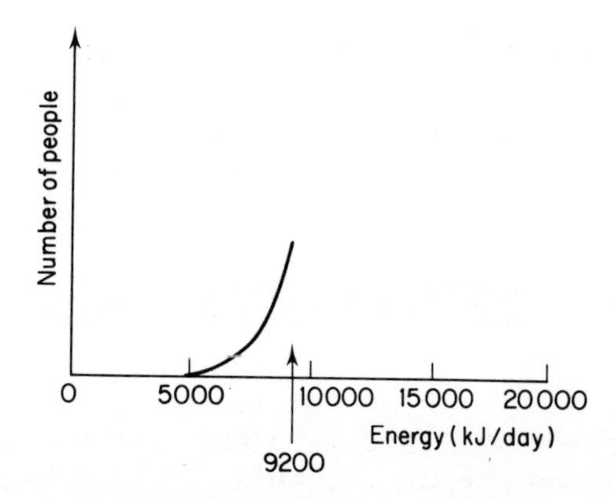

(c) There is protein in rice. 5% of the rough rice eaten in the Far East may be protein. Refined white rice usually eaten in Britain may contain less. The recommended intake of protein is 70 g/day.
   (i) What is the mass of protein in the rice eaten per day by an average person in the Far East? *(2)*
   (ii) Extra protein could be taken in by eating other foods. Name three foods which contain a lot of protein. *(3)*

(d) Brown rice contains thiamine. This is a vitamin B. Lack of thiamine causes beri beri. It produces weakening of the muscles. The person may become paralysed and die. 76% of the thiamine is removed when brown rice is 'polished' to make white rice. The bran which is removed is used for animal feed. There is a possible problem in places where the diet is mostly white rice. Explain two ways in which this could be overcome. *(3)*

*26 Electric shocks

The human body is like a bag of salty water. It conducts electricity fairly easily. But between one finger and another on the same hand the resistance may be as much as 80 000 Ω. The resistance between the armpits may be only 5000 Ω. The resistance is across the skin. With resistance-reducing cream, the resistance of the body can be lowered to below 100 Ω.

Currents above 1 mA cause a tingling feeling. Currents of over 10 mA cause muscular contraction. The person affected may be unable to let go. Above 15 mA breathing and heartbeat might be affected. Whether or not heart and lungs are affected will depend on whether or not the current passes through these organs.

(a) (i) What is the lowest current which you would expect to pass between two fingers on the same hand if they were touching the live and neutral of a 240 V supply?

$$\text{Resistance} = \frac{\text{voltage}}{\text{current}}$$ *(3)*

   (ii) Would the current passing between the two thumbs be much different from the current passing between two fingers on the same hand? Explain your answer. *(2)*
   (iii) In what circumstances would a greater current pass between two fingers of the same hand? *(1)*
   (iv) The electric shock caused by 240 V between the armpits would be more serious than that caused by 240 V between two fingers of the same hand. Give two reasons for this. *(2)*
   (v) Electricians who work on 'live' equipment wear rubber boots. They are advised to keep one hand in their pocket. Explain this. *(2)*
   (vi) Birds perch on 400 kV pylon lines. They do not get electric shocks. Explain this. *(1)*
   (vii) Surgeons must be very careful not to give patients small electric shocks during operations. An instrument working from a small battery could give electric shocks too small for the surgeon to feel them. But these could kill a patient. Explain this. *(4)*

(b) The graph shows the current at which half the men or women tested could not let go of electrodes.

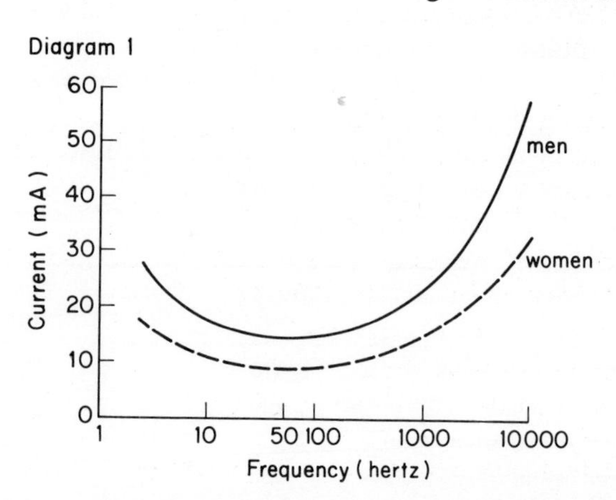

   (i) The mains frequency in Britain is 50 Hz. Is this the best frequency from a safety point of view? Explain your answer. *(2)*
   (ii) Give possible explanations for the difference in the graphs for men and women. *(3)*

(c) The inside of a nerve fibre is usually negative compared with the surrounding tissue. This is because it contains fewer sodium ions (Na^+). When electrodes are placed on the skin, a current flows through the nerve fibre. This causes a movement of sodium ions through the nerve membrane. Examine the diagrams and explain what must happen to the sodium ions. (2)

Diagram 2

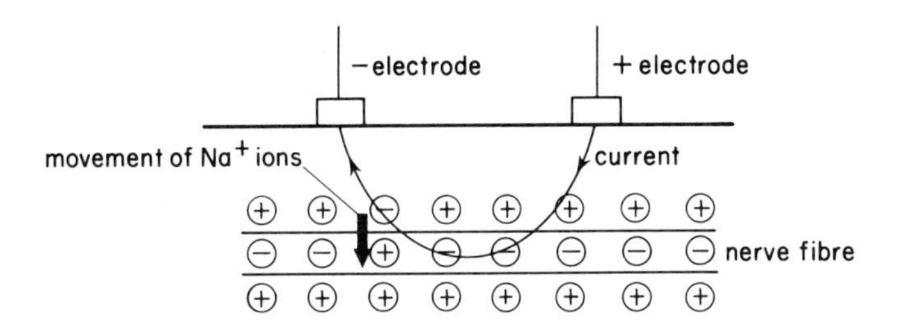

*27

Read the statements below and use them to construct a flow diagram (or spider diagram) to show the relationships of the body's carbohydrate and fat pool to different parts of body metabolism.

Carbohydrate and fat metabolism

(a) Fats and carbohydrates are interchangeable and can be considered together for this flow diagram.

Fat $\rightleftharpoons$ carbohydrate

(b) Fats and carbohydrates form a central pool with at least eight metabolic links.
(c) Fats and carbohydrates enter the body by ingestion and then digestion.
(d) A very small amount of carbohydrate is lost by excretion via the kidney.
(e) Fats are used to make complex molecules such as steroids and acetylcholine.
(f) Excess fat is stored.
(g) Excess carbohydrate is stored as glycogen or fat.
(h) Carbohydrate and fats, when changed into organic acids and combined with ammonium ions (NH_4^+) can form amino acids.
(i) Excess protein is changed into amino acids, ammonium ions are removed from them and the remaining acid enters the carbohydrate/fat pool.
(j) Most carbohydrate is respired by body cells to produce carbon dioxide and water as by-products of energy release.
(k) Some fat is secreted as oil on the skin surface.

*28 Transmitting electricity

Electricity transmission cables are usually made from pure metals. The data below are for wires with a cross-sectional area of 1 cm^2 (0.0001 m^2).

Metal	*Resistance* (Ω/km)	*Density* (kg/m^3)	*Strength*
aluminium	0.28	2700	low
copper	0.17	8800	medium
iron	1.00	7900	high
magnesium	0.39	1740	medium
nickel	0.68	8900	high
zinc	0.59	7100	medium

(a) (i) Which metal from the above list would you choose for underground electricity transmission cables? (1)
(ii) Give a reason for your choice. (1)

(b) For overhead transmission cables the choice of metal depends on the mass in kilograms per ohm of resistance.
(i) Suggest possible reasons why mass is also considered. (3)
(ii) For the above metals, calculate the mass (kg) per ohm. (6)
(iii) Using your results, select the most suitable metal for overhead transmission cables. (1)
(iv) Are there any disadvantages in using the metal you have chosen? Explain your answer. (2)

(c) In practice, most overhead cables have a thin iron core which is surrounded by aluminium.
(i) Explain why this procedure is used. (5)
(ii) What is the resistance of this type of cable? Assume that the iron has a resistance of 5 Ω/km and that aluminium is 0.28 Ω/km. (3)

(d) Overhead cables can become covered by ice in winter. The extra weight of the ice can damage them. The normal transmission voltage may be 132 kV. This means that to transmit 1 MW of power a current of only 7.57 A is needed. The Electricity Board can remove the ice by temporarily disconnecting the cables from the National Grid and changing the current. Explain this.

*29 Fossil fight

These fossil remains were found in a rock bed. Two scientists tried to reassemble them. They were allowed to use plastic filler to make up for missing bones. Both made different reconstructions and both were sure that theirs was the most accurate.

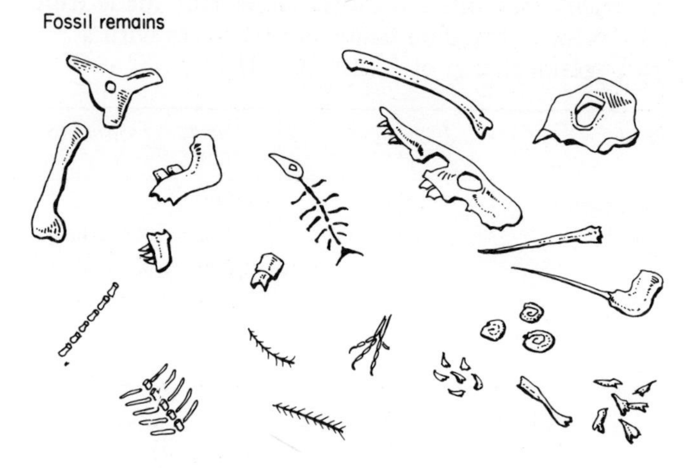

Draw both reconstructed fossils and give five good reasons why one is the most accurate. Think about movement, feeding, balance and anything else you can tell from a fossil.

30 Testing milk

Untreated milk always contains an enzyme called phosphatase. It is quite easy to test for it. Some raw milk may contain tubercle bacteria. These cause tuberculosis. They may be present in such small quantities that they are not easy to detect. Milk is pasteurized by keeping it at a high temperature for a certain length of time. This kills some living bacteria and makes some enzymes inactive. The graphs show the 'death curves' for tubercle bacteria and phosphatase. In zone A, the bacteria are alive and phosphatase is active. In zone B, the tubercle bacteria have been killed but the phosphatase remains active. In zone C, the bacteria have been killed and phosphatase is inactive.

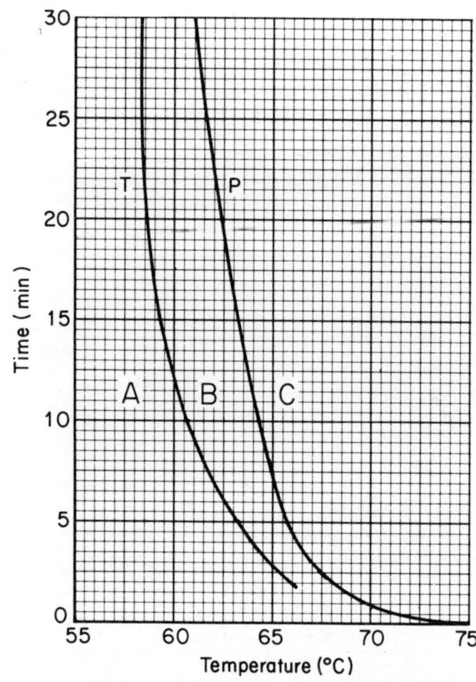

(a) Which of the two things for which 'death curves' are drawn is most easily made inactive? Explain your answer. *(3)*

(b) Only 0.001 cm^3 of milk can be examined on a microscope slide. This presents a problem.

  (i) If a litre (1000 cm^3) of milk contains 2000 tubercle bacteria, how many are there likely to be in 1 cm^3 of milk? *(1)*

  (ii) How many bacteria are likely to be found when 2000 microscope slides of the milk are examined? *(2)*

  (iii) Why is this method for finding tubercle bacteria unsatisfactory? *(2)*

(c) (i) Use the graphs to find the minimum time for which milk should be held at 65°C to kill tubercle bacteria. *(1)*

  (ii) It is not easy to ensure that all the milk has been at this temperature all the time. Explain fully how environmental health inspectors could ensure that a pasteurization process operating at 65°C would be killing all tubercle bacteria. *(3)*

(d) Milk is pasteurized to kill harmful bacteria. It used to be pasteurized at 62°C for 30 minutes. It is now pasteurized by a high-temperature short-time (HTST) method at 72°C for 15 seconds.

  (i) What can you say about the safety of
    the old method, *(1)*
    the new method? *(1)*

  (ii) Compare the likely cost of the two methods. *(2)*

*31 Rusting

A pupil knew that oxygen and water had to be present for iron and steel to corrode, but wanted to know how other substances and temperature affected the amount of corrosion. The pupil performed a series of experiments varying the liquids and temperatures used. In these experiments nails of similar size were used, the volume of liquid was kept constant and the length of time for each experiment was the same.

The following procedure was used for each experiment.

1. The nails were cleaned with emery paper and weighed.
2. The experiment was set up and left.
3. After the required time, the nails were removed. Any corrosion was cleaned off and then the nails were dried and reweighed.
4. The percentage loss in mass was calculated.

(a) Why did the pupil use nails of about the same size? *(2)*

(b) Why was the volume kept the same for each experiment? *(1)*

(c) Why was the same time used for each experiment? *(1)*

(d) What else, not mentioned in the description of the experiment, should also be kept constant? *(1)*

(e) (i) Why was the percentage loss in mass calculated rather than just giving the loss in mass? *(1)*

(ii) How would you calculate the percentage loss in mass? *(2)*

The table below shows the percentage loss in mass using various liquids at different temperatures.

Liquid	*Temperature* (°C)						
	10	20	30	40	50	60	70
distilled water	4	5	7	10	14	19	25
sea water	10	12	16	22	30	40	52
dilute acid	15	20	30	50	90	100	100
dilute alkali	9	10	13	18	25	34	45
concentrated nitric acid	1	1	2	2	3	3	3
alcohol solution	2	3	5	7	10	13	17

(f) There is a pattern between the amount of corrosion and temperature. Describe it. *(3)*

(g) There is a pattern between the liquids used and the amount of corrosion. Describe it. *(3)*

(h) (i) Which liquid gives an unexpected result? *(1)*

(ii) Suggest a possible cause for this result. *(2)*

(i) Suggest a possible reason why the alcohol solution causes less corrosion than distilled water. *(2)*

(j) (i) What additional process occurs when iron is added to dilute acid? *(4)*

(ii) Write the word equation for the process, choosing your own acid. *(1)*

(iii) Write the balanced formula equation for (j)(ii). *(2)*

32 Autoclave

In hospitals some equipment is heated in an autoclave before use. An autoclave is a kind of pressure cooker. The equipment has to be heated to a temperature of 120°C. Diagram 1 shows an autoclave.

Diagram 1

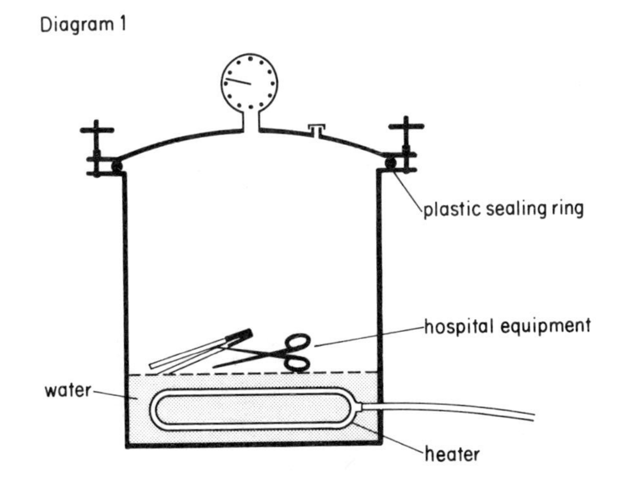

The table gives some information about the boiling points of water and the pressure needed.

Boiling point of water (°C)	*Pressure* (kPa)
80	49
90	68
100	99
110	140
120	195
130	273

(a) (i) Why must some hospital equipment be heated to 120°C? *(1)*

(ii) Complete this sentence: The nurse can turn off the autoclave when the pressure gauge reads at least . . . . kPa. *(1)*

(b) The graph shows how atmospheric pressure changes with height above sea level.

Diagram 2

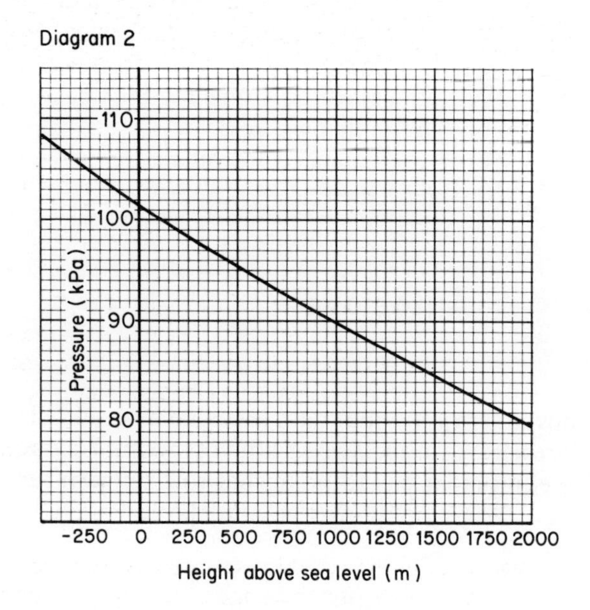

(i) What is the pressure at sea level? *(1)*

(ii) Where would atmospheric pressure be 104 kPa? *(1)*

(iii) What is atmospheric pressure at a height of 1500 m? *(1)*

(c) Kaimosi hospital in Kenya is at a height of 1500 m. If water is boiled in an open pan at Kaimosi it does not boil at 100°C. Use the table and the graph to estimate its boiling point. *(2)*

(d) London is at sea level. Use the graph to work out atmospheric pressure in London on a normal day. *(1)*

(e) What is the difference in pressure between the inside and outside of the autoclave at 120°C:

(i) when it is used in London, *(1)*

(ii) when it is used in Kaimosi? *(1)*

(f) The plastic sealing rings eventually perish. They must be replaced frequently. If they are not replaced there could be accidents. This is because of the big pressure differences between the inside and outside of the autoclave. There are more accidents with an autoclave in Kaimosi than in London. Use your answers to (e)(i) and (ii) above to explain why. *(1)*

(g) Explain a method which could be used in Kaimosi to make hospital equipment safe for use without an autoclave. *(2)*

(h) You must design a special autoclave which is safe to use at places like Kaimosi. The pressure between one side of a sealing ring and the other must never be more than 100 kPa. But it must be possible to reach a temperature of 120°C. Draw a labelled diagram of your special autoclave. Describe how it works. *(5)*

Talking about the answers

1 Information handling

Information can be given in a table, a graph, a pie diagram, a bar chart, a labelled diagram, a few paragraphs of writing, and in many other ways.

You may have to change information from one form to another: from a table of numbers to a graph or from a paragraph of writing to a pie diagram.

You may have to collect together information from a number of different sources. For this you may have to decide what kind of information is relevant before selecting pieces of data.

Sometimes information has to be processed. Numbers have to be added, subtracted and dealt with in other ways. This requires some skill in arithmetic. You have to know what needs to be done and you must be able to do it correctly.

1 This question is about the properties of materials. It concerns decisions about a technological process. You have to select the piece of information which is most important for this particular purpose.

Being easy to shape **E** is an advantage. Being difficult to colour **B** may be a slight disadvantage if people want coloured plates. Paper plates would be biodegradable. Plastic ones would not be. Biodegradable things are easier to dispose of. But not being biodegradable **D** is only a slight disadvantage. Cracking when hit by a spoon **C** would be a serious disadvantage if users were going to do a lot of hitting with spoons. But hitting with a spoon should be accidental and unintended. Boiled food from water at 100°C or fried food from fat at even higher temperatures *must* be put on plates. So becoming flexible **A** at 100°C is the *main* disadvantage.

Answer **A**

17 (a) This asks you to change information from one form to another. Six columns must be completed correctly for the full three marks. One mark is taken off for each mistake. So four correct columns would get just one mark. Only three columns correct gets zero. Being right only half of the time is not good enough (The plotting of points for graphs is often marked in this way too.) *(3)*

(b) Forearm *(1)*

(c) Wrist *(1)*

It is very easy to see which bones were broken most and least by looking at the bar chart. But you may have made a mistake in completing the bar chart. So it is a good idea to check with the table as well.

(d) 37. In data handling you will often have to do arithmetic. A calculator can be used, but you can easily miss a number. It is good practice to add twice: 12 + 2 + 1, etc, and 4 + 6 + 5, etc, to check your answer. *(1)*

(e) This section is about interpreting data accurately without jumping to conclusions. At least one person had broken more than one bone.

Avoid making statements about which you cannot be sure, such as "one person had broken three bones" or "two people had broken two bones". But you can say, "*Perhaps* one person had broken three bones". *(1)*

The question could have asked for three possible explanations (for three marks). Can you think of three?

(f) This asks you to make a judgement. Can we rely on the information in order to make statements about the whole school? Remember the information is only about 35 first years who had broken bones. It does not tell us whether *only* 35 first years had broken bones. It does not tell us what percentage of first years had broken bones. But it does compare breakages of toes with wrists, and it says something about the number who had broken more than one bone.

There are two marks, so we are looking for two things: a comment which is (1) valid (true) and which also (2) supports your judgement. Let us look at some possible correct answers.

(i) Yes. Thirty-five is a big sample. There might be only 175 pupils who have broken bones in the whole school, so we have information about ⅕ of them.

(ii) No. Older pupils are bigger and stronger than first years. They might break different bones.

(iii) Yes. These figures are for breakages up to the age of 11½. Secondary pupils do not often break bones. So the figures for pupils of 14½ will not be much different.

(iv) No. The number of people who have broken more than one bone is bound to increase as we go from first years to fifth years. *(2)*

So there is more than one possible answer. The question could have said, 'Discuss whether or not these results fairly represent those for the whole school.' This could have been for six marks. How would you have answered it? There are a lot of questions like part (f) in science. For example, 'There are six daisies and four dandelions in 1 m^2 of the playing field outside the lab. Is this representative of the whole field?'

Parts (e) and (f) were asking for conclusions and judgements. There are more questions of this type in the chapter called *Evaluation*.

18 This question asks you to select and bring together information from two sources. Put your ruler on the graph along the line for intensity = 1.7 W/m^2. There are two sections below your ruler. There is not enough light between:

(i) 1.7 m and 3.9 m from end A,
(ii) 7.3 m from end A and the opposite end of the bench.

Go back to the first diagram. The people working in these sections are:

(i) Miss Webb,
(ii) Mrs Gutowski.

We note that Miss Jones at 4 m from end A has only 1.75 W/m^2. But the rule says, 'not less than 1.7 W/m^2'. So she does not need more light.

19 The diagram gives enough information for you to be able to understand how the device works. You have to put this information down in words. The question is about changing information from one form to another.

It is sometimes easiest to put down a series of numbered statements. There are two parts to the device: (a) the pivoted lever and (b) the paper on the rotating arm.

(a)
(i) As the man breathes in, the band at the front of his chest moves upwards. *(1)*
(ii) This makes the lever rotate clockwise on the pivot and the pen moves downwards. *(1)*
(iii) As the man breathes out, the band at the front of his chest moves downwards.
(iv) This makes the lever rotate anticlockwise on the pivot and the pen moves upwards.
(1 *mark for the opposite of what we had in (i) and (ii)*)

(b)
(v) A motor rotates the drum and the paper which is fastened to it.
(vi) As the paper rotates, the pen draws a line on it.
(vii) Because the pen moves up and down as the man breathes, the pen draws a wavy line.
(1 *mark for an explanation of why the pen draws a wavy line, i.e. (v), (vi) and (vii) total 1 mark*)
(viii) Breathing rate is equal to the number of waves drawn per minute. *(1)*
(ix) X shows the position of the pen when the man has breathed out. *(1)*
(x) Y shows the position of the pen when the man has breathed in. *(1)*

26 The way information about light and carbon dioxide in the lake is presented is unusual. It tells us that there is plenty of oxygen at the top of the lake and that it decreases until there is none at the bottom of the lake. What about carbon dioxide? You need to bring together information from the table and the first diagram and apply this to the diagram of the lake.

In real situations there is usually a lot of information available. We need to select only some of it. In this case the data about things produced by photosynthesis are not needed.

(a)

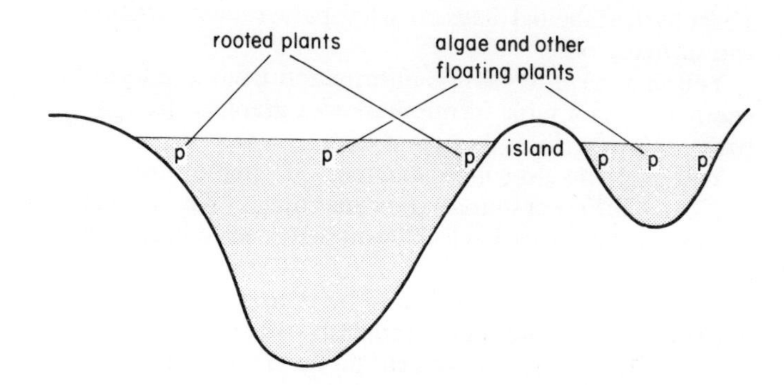

(b) At the surface there is plenty of light *(1)* and carbon dioxide. *(1)*
At the bottom there is enough carbon dioxide but not enough light. *(1)*
At middle depth there is not much carbon dioxide and not much light. *(1)*

The information about light and carbon dioxide will be different for different lakes. You could extend this question by considering the carbon dioxide. Where in this lake may animals be producing it? Where may decay processes be producing it? What makes these decay processes take place? How do pressure and temperature affect the amount of carbon dioxide dissolved in the water?

You could do the same sort of thing for light. How might acid rain or other pollution affect these things?

This is a series of problems. To solve them you would need to look up data in textbooks or an encyclopaedia. You would look for patterns (how the solubility of carbon dioxide varies with temperature, how temperature in a lake varies with depth). You would need to make an evaluation of different patterns: one factor makes the amount of carbon dioxide increase with depth, another makes it decrease with depth.

Bringing all these skills together is called synthesis. There are questions of this type in a later chapter.

35 You are given information in a table. You are asked to change this to a graph. You have to select pieces of information from the table or the graph and you have to interpret some of the information.

(a) 5 marks: subtract 1 for each error. The line for hydroelectricity is almost on the horizontal axis.
(b) (i) Coal (1)
(ii) Any sensible reason, such as caused pollution/bulky and dirty to handle/more convenient fuels became available. (1)
(c) (i) Natural gas (1)
(ii) Any sensible reason, for example, it only became available in the 1960s/safe and convenient to use. (1)
(d) (i) Hydroelectric (1)
(ii) Any sensible reason, such as limited to mountain areas/expensive to build/all possible sites used up. (1)
(e) (i) 267 (1)
(ii) 327 (Remember it is good practice to check arithmetic) (1)
(iii) Increase of 60 million tonnes (1)
(iv) Any sensible reason, for example, more people/more energy being used by industry/more energy being used for domestic heating. (1)

40 (a) (i) SNPS or ethylene glycol (1)
(ii) SBN (1)
(b) Car anti-freeze (1)
(c) To prevent the material of the radiator corroding. (1)
(d) To ensure that the inhibitor did reduce the corrosion. (1)
(e) Aluminium forms an oxide film on its surface which hinders further corrosion. (1)
SBN (1)
(f) pH less than 7 is acidic. (1)
Acids react with most metals. (1)

41 This is a comprehension question. Information is presented in two paragraphs. You have to select pieces of information.

(a) Any two of: cheeses/bread/beer/wine/animal food. (2)
(b) Alcohol. (2)
(c) Special fungi are needed for each type of cheese. (1)
Pure cultures of these are kept. (1)
(d) By using a variety of cheese fungi (moulds). (2)
(e) They can be grown in tanks of sugar waste (1)
and then harvested to provide animal food. (1)
(f)

Sun → plant → (chemical energy store (sugar)) → fungus → (chemical energy store (protein)) → animal food

(g) Will turn waste into animal food (1)
which can be sold (1)
(h) Any two of: the tank must be sterile or unwanted organisms will grow/oxygen supply/right balance of sugar and minerals. (2)
(i) Palatable for animals. (1)
Capable of being processed and balanced for animal growth. (1)
(j) Carbohydrate gives energy only. This is changed into protein which is useful for growth. (1)
Nitrogen and minerals are needed. (1)

2 Problem solving

How do you start to solve problems? Sometimes you have to begin by asking 'What exactly is the problem?' For example, you may be interested in this problem:

'Do big worms move faster than little worms?'

What does big mean? Does it mean long, fat, grown-up, heavy? What does fast mean? Does it mean distance/time averaged over a minute or the fastest sprint observed? Is it for a straight line 'walk' (for example, between two rulers) or for a random walk?

Sometimes it is obvious what the problem is. So you have to decide what is causing the problem. This can be done by matching possible causes with data or your knowledge of scientific principles.

You then have to design a solution to the problem. One of the stages may be identifying variables (things that affect the problem) and deciding how to control them. You have to have reasons for your solutions. There may be more than one solution.

Some solutions may be better than others. Different solutions can be judged. Some may not work. One may be quickest, one may be cheapest, one may be the most effective for removing the problem. This is evaluation. You will have opportunity to do more evaluation in a later chapter.

Sometimes the particular problem is only one example of a general one. What current I will flow through a resistor R when the voltage is V? The general solution is $I = V/R$. All you have to do is put the right numbers into the formula.

1 (a) What is causing the problem? There are two possibilities: first that there is something outside that ants do not like, second that there is something inside that ants do like. Mrs McGrath thinks of four things which might affect ants. These have to be matched with the data about what ants like and do not like.

A. Warmer outside. Ants like warmth so this would make them stay outside.
D. Paraffin in the kitchen. Ants do not like paraffin so this would make them stay outside.
C. Cream cake in the freezer. Ants eat cake. But would they know that it was in the freezer? Ants like warmth so they would be unlikely to be tempted into the freezer.

B. Sugar on floor. Ants eat sugar. They could get at it easily. This has probably made the ants come in. Answer **B**

(b) Mrs McGrath will need to deal with the ants which are already in the kitchen. But first she must stop any more coming in. How can the problem be solved? Which of the possible solutions is supported by a good reason?

A. Paraffin to garage. Ants do not like paraffin. This could make any ants in the garage move out, perhaps into the kitchen.

D. Heating on. Ants like warm places so this could bring more ants into the kitchen.

B. Sugar across doorway. Ants would be attracted to the sugar from outside and inside the house. Perhaps some more from outside would come in. It is difficult to be sure. (You could try it to find out.)

C. Cinders (ashes from a fire) across doorway. Ants do not like cinders so this should stop them coming in. Answer **C.**

6 All of the substances listed would react with acid. But the water must be made slightly alkaline (pH 8–9) not just neutral.

B. Calcium carbonate would neutralize the acid but the pH would then be 7. Calcium carbonate does not dissolve in neutral water.

A. Ammonia solution could be added until the pH became 8 or 9, but any extra ammonia solution would certainly dissolve in water. The pH would then be higher than 9. The water could give off ammonia gas.

C. Sodium hydroxide solution could bring the pH to 9 but as with ammonia solution, if too much is added, the pH would become higher than 9.

D. Magnesium hydroxide would remove the acidity. Because it is very slightly soluble in water it will give a weak alkaline solution of about pH 8–9.

Answer **D**

8 (a) This part of the problem is to design an experiment. You have to decide what variables might affect the result. These have to be controlled (kept the same in each test). Here are some of the things that might affect the result:

quantity of fabric
quantity of dye
quantity of water
temperature
washing process
the original colour of the material.

There is one mark for each of three correct statements. *(3)*

(b) What is the problem? Which dye produces exactly the same colour as shown on the packet? (Compare for colour match.) Which dye produces the darkest colour? (Measure amount of light reflected from the fabric using a light meter.) *(1)*

(c) It is good practice in any experiment to repeat the test several times. This makes sure that the results are reliable. *(1)*

18 With question 8 and some of the questions that followed it, there seemed to be a pattern for designing an experiment. You had to think what variables could affect the result and these had to be kept constant. You also had to decide what really could be measured to get a result from the experiment. Using the same pattern helped you to solve a number of problems in designing experiments. But you need to know how to apply the pattern in different situations.

This question gives a pattern as a formula. But you have to know how to use the formula in different situations. The rule for using the formula is that the distance considered must be in the same direction as the force. The force is the weight of the bricks (200 N). It is a vertical force. The vertical distance between X and Y is 4 m. This is the distance which we must use in the formula.

Answer $200 \times 4 = 800$ J

31 The passage gives a design brief for four devices. The paragraph about the cooker gives many precise details about design requirements. The paragraph about hairdressing gives fewer details. You may be able to think of some additional things that would have to be controlled in hairdressing. This is another problem in which you have to identify variables.

For each device you must draw *(1)* and label *(1)* a control panel and it must have a sensible design *(1)* *(3)*

The cooker would need controls for

(a) Oven: temperature, time meal is to be ready, how long it takes to cook.
(b) Hopper control: quantity from each hopper.
(c) Type of mixing.
(d) Removal from freezer and fridge.
(e) Time for thawing.
(f) Microwave: time meal is to be ready, cooking time. *(6)*

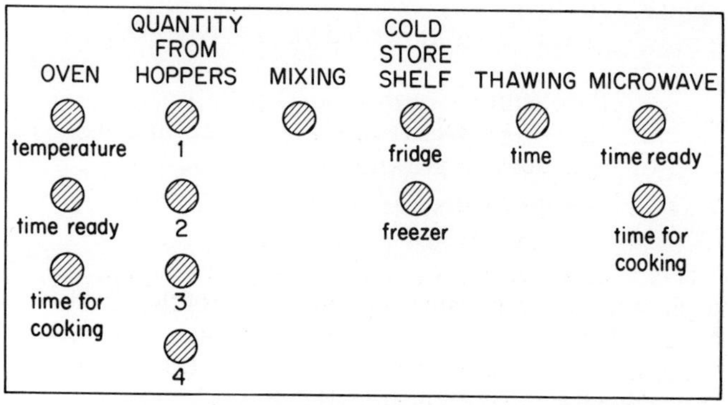

OVEN	QUANTITY FROM HOPPERS	MIXING	COLD STORE SHELF	THAWING	MICROWAVE
temperature	1		fridge	time	time ready
time ready	2		freezer		time for cooking
time for cooking	3				
	4				

33 (a) He could have two possible problems. They are clearly stated. Either he has been driving with more than 80 mg of alcohol per 100 cm^3 of blood *(1)*

or he may have been driving too fast. *(1)*

The numbers allow us to test these possibilities. If he drove at the speed limit of 70 m.p.h., it would have taken him 2 h to travel 140 miles from Lancaster. *(1)*
Two hours earlier he would have had

$55 + (2 \times 15) = 85$ mg of alcohol/100 cm^3 of blood. *(1)*

If he left Lancaster when his alcohol level was just on the limit, how long would his journey have taken?

$80 - 55 = 25$ mg of alcohol/100 cm^3 of blood.

His body would have disposed of 25 units of alcohol. *(1)*
This takes

$\frac{25}{15} = 1.67$ h

So his speed would have been

$\frac{140}{1.67} = 84$ m.p.h.

(In these calculations you are using the formula for speed and general ideas about proportionality.)

(b) An hour earlier his blood alcohol level may have been

$55 + 15 = 70$ units *(1)*

This is still legal. If he drank alcohol at this time, his blood alcohol level prior to this would have been lower than 70 units. *(1)*
So he could argue that he was within the alcohol limit all the way from Lancaster.

(c) In the rest of the question you need to link the information given to relevant facts about the way the body works.

Alcohol is a small molecule and it is absorbed into the blood through the wall of the stomach. The rate of absorption may depend on whether the stomach is empty or full of other food. Alcohol is broken down by the liver. The carbohydrate from alcohol gives energy for doing work. Doing a lot of work will increase the metabolic rate. More carbohydrate will be needed and alcohol will be broken down faster. So the rate at which alcohol enters and leaves the blood is variable.

Water enters the blood through the wall of the colon. It is removed from the blood by the kidneys, goes to the bladder and is passed in urine. If the two processes do not take place at exactly the same rate, the volume of the blood may vary. If there is a lot of salt in the blood, the kidneys will extract water more slowly so that the salt in the blood does not become too concentrated. If there is a lot of water in the blood, the alcohol is more dilute.

(i) 3–5 litres *(1)*
(ii) water, salt, urea (mostly) *(1)*
(iii) There is really no answer to this problem because there are too many variables which cannot be controlled. For example:
The other contents of the stomach, the level of salt in the blood, the amount of exercise being done. *(1)*
The more concentrated alcohol in whisky may be absorbed more quickly by the wall of the stomach if the stomach is empty. *(1)*
The greater quantity of water in beer may temporarily increase the volume of the blood, so diluting the alcohol. *(1)*

(d) (i) You should be familiar with the pattern for working out concentration:

$$\text{Concentration} = \left(\begin{matrix}\text{volume of}\\ \text{solute}\end{matrix}\right) \div \left(\begin{matrix}\text{volume of}\\ \text{solution}\end{matrix}\right)$$

In (c)(i) we gave the volume of the blood as 3–5 litres. Let us take the volume of the blood with alcohol in it as 4 litres. The concentration is

$\frac{50}{4000} = 0.0125$ or 12.5 parts per 1000. *(1)*

(ii) If all the alcohol in 1 litre of beer is taken into the blood, this would be 12.5 times higher than the legal limit. *(1)*
So only a small part of the alcohol consumed must go into the blood. *(1)*

(e) The metabolic rate must be increased (that is, the rate at which the liver processes alcohol). *(1)*
This will happen if more energy is needed, for example, because the man is running. *(1)*

3 Patterns

Some of the questions in this chapter are about finding patterns. You have to build up information to make a pattern. You may have to describe the patterns.

Once you have seen what the pattern is, you can use it. You can recognize pieces of information which do not fit the pattern. You can link together pieces of information which do fit the pattern. You can make predictions.

2 You must find the pattern and then decide which of the possible answers describes it best.

Consider the effect of temperature increase when pressure is constant. Look down the column of figures for pressure = 25 atmospheres. The volume of ammonia produced seems to get smaller and smaller as you go down the column. Is the pattern the same for every column of figures? Yes. But as we look down a column of figures, temperature increases from 100°C to 500°C. So the amount of ammonia decreases as the temperature increases. This is the same as saying: 'The amount of ammonia increases as the temperature decreases'.

We know that answers **A** and **B** are wrong; **C** or **D** may be correct.

Now look along the row of figures for temperature = 100°C. The figures increase as we look from left to right and this pattern is repeated in every row. As we look along a row of figures, the pressure increases from 25 atmospheres to 400 atmospheres. We can say, 'The amount of ammonia produced increases as the pressure increases'.

The correct description of the pattern is **C**: 'The amount of ammonia increases as the temperature decreases and the pressure increases'.

12 (a) Most of the smells cannot be detected by everyone. The exception to this pattern is the smell of sweaty socks. Everyone in the test could smell them from a distance of 10 m. *(1)*

(b) Patterns are not always perfect. You can say that, in general, people aged 41–60 cannot detect smells as well as younger people. (Sweaty socks are an exception) *(1)*

You cannot really find any difference between the 11–20 and the 21–40 age groups. For coffee and curry the 11–20 year olds do best. For onions and perfume the 21–40 year olds do best. The differences are small. The ability to smell seems about the same for 11–20 year olds and 21–40 year olds. *(1)*

15 In science you must be able to recognize patterns even when the information is not arranged in the proper order. It is easiest to see what is happening if you think of the third lens (thinnest) and its cross (smallest) as coming before the other three.

(a) The fatter the lens the bigger the cross. *(1)*
But can you be more precise? So far as you can see from the diagrams, if the fatness of the lens doubles, the size of the cross doubles. In fact the width of the cross seems to be the same as the fatness of the lens producing it. This is not quite correct. A 'lens grinder's formula' gives a slightly different result. But you can say:

The size of the cross is approximately proportional to the fatness of the lens. *(1)*

There is another point. What is meant by the size of the cross?

The height, width and thickness of the cross are always magnified by the same amount. *(1)*

(b) You have to use the pattern to make a prediction. The lens is about half as fat as the thinnest of the ones in the first set of diagrams. So the cross seen through it will be about half as big. *(1)*

(c) Graph (3); 'interpolation' to get size of cross. *(1)*

23 (a) You have to build up a pattern on graph paper. It is not a graph in the usual sense. Some of the information has been plotted in the graph. You can do the rest of it yourself. The marking here requires correct axes, scales and plotting. A mark is subtracted for each error. *(5)*

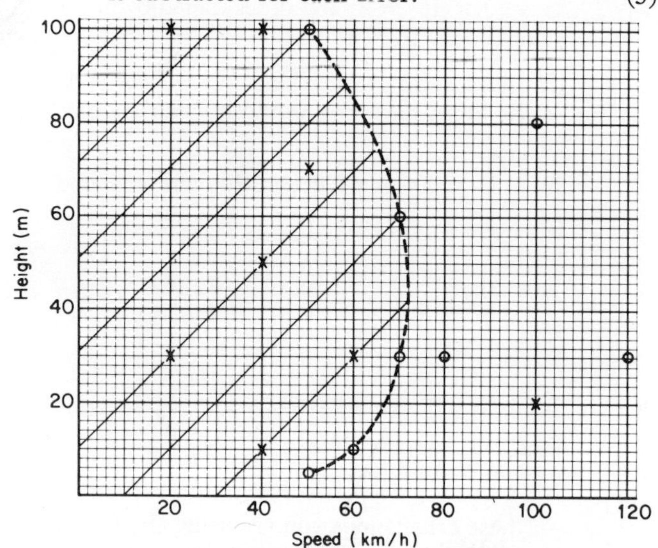

(b) A curve has been drawn linking cases of helicopters which did not crash. All cases to the left of this curve crashed. So the curve is one limit of safe operating conditions. You cannot go beyond the 5 m, 50 km/h point because you do not know what happens. It would be an unsafe prediction. Idea that known safe landings rather than known crashes define the limit of safe operation *(1)*
Two correct curves *(2)*
No extrapolation beyond known points *(1)*
Shading *(1)*

(c) This calls for use of the pattern you have built up.
(i) One limit seems to be about 6 m.
What is the other limit? *(2)*
(ii) Its minimum safe speed is about 72 km/h. *(1)*

30

Distance d (cm)	*Sag s* (mm)	*Increase in sag* (mm)
20	0.8	
25	1.6	0.8
30	2.7	1.1
35	5.9	3.2*
40	6.4	0.5*
45	9.1	2.7
50	12.5	3.4
55	16.6	4.1

It is a good idea to arrange data in 'rank order'. This means putting down values for distance *d* as 20, 25, 30, 35, 40, etc.

(a) 2 marks for rank ordering. (Subtract 1 for each error) *(2)*

The matching data for sag also increases. So there is some kind of pattern. One way of sorting out which piece of information is wrong is to look at the increase in sag between one line and the next. The increase seems to get bigger all the way down. There seems to be something odd at *. The 3.2 mm increase in sag is too big (similar to the increase when distance goes from 45 cm to 50 cm.) The 0.5 mm increase in sag is too small (similar to the increase when distance goes from 20 cm to 25 cm.) So the value of 5.9 mm is probably wrong.

(b) 2 marks for identifying this. *(2)*

(c) 1 mark for explaining that the difference in sag should increase all the way down the column. (You can get this mark even if (b) above is wrong.)
1 mark for stating that the 3.2 mm increase is too big.
1 mark for stating that the 0.5 mm increase is too small. *(3)*

33 The gases are listed in alphabetical order. When you first look at the data, there seems to be no pattern. But parts (a) and (b) help you to find it.

(a) $32 \times 0.5 = 16$. The gas is methane. *(1)*

(b) $\frac{0.72}{1.44} = \frac{1}{2}$ *(1)*

(c) You have seen from (a) and (b) that when the relative molar mass is halved, the density is also halved. The two sets of numbers may be proportional to one another. If you add together the figures for carbon monoxide and methane, you get the figures for carbon dioxide. Some people would be able to recognize this as proportionality. You could confirm this by drawing a graph.

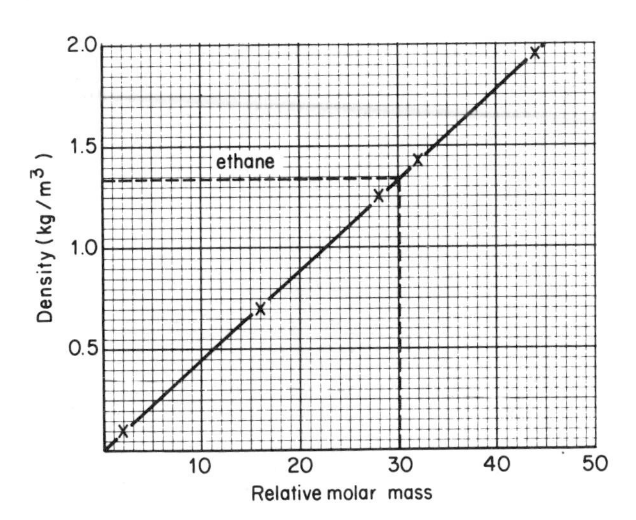

Density is proportional to relative molar mass. *(1)* The graph could be used to find the density of ethane, or you could do a calculation.

Density = $C \times$ (relative molar mass)

Where C is a constant, we can find C by putting in numbers for any of the gases. Try carbon dioxide.

$1.98 = C \times 44$

Therefore $C = \frac{1.98}{44} = 0.045$

Now we have the pattern as a formula,

Density = $0.045 \times$ (relative molar mass)

For ethane

Density $= 0.045 \times 30$
$= 1.35$ kg/m^3 *(1)*

4 Evaluation

Here you may have to draw conclusions from data. The data can be presented in a variety of ways, for example, as diagrams, tables or paragraphs of writing. You must be able to make judgements about information. How reliable is it? What exactly does it mean and what are the possible incorrect interpretations of the information? You will be making decisions based on the examination of the evidence. Sometimes you may feel that essential evidence is missing and you will have to identify it.

Statements and explanations about unfamiliar effects have to be assessed. Are they correctly based on scientific principles? Have some science principles been over-emphasized and others overlooked?

Experimental procedures can be criticized. Have all variables been controlled? Are measurements reliable? Have proper safety precautions been taken? What do the results really indicate? What improvements could be made to the experiment?

Solutions to problems, such as devices intended to control heating, car braking systems, must also be evaluated.

3 Biopol can be broken down by bacteria. For some things this is an advantage. But for others it could be a disaster. Biodegradable plastic pants for babies would not last long!

A. Capsules for medicines. The capsules must be broken down in order to release the medicine inside the body. So Biopol would be suitable.

C. Pots for plants. Bacteria in the soil would digest Biopol pots. The plant would no longer have its roots restricted by the pot. They could spread out into the surrounding soil. This would be very useful. If the plant stood in its pot for too long before planting in the ground, the pot could start to deteriorate. But plants should not be kept like this. On balance, Biopol pots would probably be an advantage.

D. Crisp packets. Presumably there are no bacteria inside the packet before it is opened. So a sealed packet made of Biopol should not deteriorate. There could be a problem if the packets are stored in a place where there are bacteria. But food should not be stored in such places. After the crisps are eaten, the packet will be rubbish and could be blown or thrown away as litter. Being biodegradable, it will eventually rot away to nothing. So Biopol would be useful for crisp packets.

B. False teeth. There are bacteria in the mouth. Some of them are in plaque. They contribute to gum disease and tooth decay. False teeth must last for many years. Biopol teeth would be very unsatisfactory. You might wake up one morning to find that you had digested your bottom set!

Answer **B**

9 The four possible sites must be evaluated. You really need more information in order to make a good decision. The map gives you data about the availability of coal and water only.

(a) There must be coal and fresh water.
- **B.** Coal and seawater but no fresh water, so unsuitable.
- **D.** Fresh water and seawater but no coal, so unsuitable.
- **A.** Fresh water available and coal nearby. A possible site, but not ideal.
- **C.** Coal and fresh water available on site. So this is probably the most suitable.

(b) Other factors have to be considered in selecting a site.
- **B.** The wind would usually blow over the agricultural land before reaching the power station. Acid rain would affect land on the opposite side of the power station. Statement (**B**) is irrelevant.
- **C.** Transport for getting building materials to the site and for getting ash away from the site may be a problem, but boats are not essential. Road and rail transport could be used for building materials. The fuel, coal, is of course on site and need not be transported.
- **A.** Very cold weather and high winds could make it difficult to build a power station. But these problems can usually be overcome. In any case, these weather conditions would create a demand for electricity in the area for heating, etc.
- **D.** Electrical energy is converted into heat energy as current passes through cables. If there is a large distance between a power station and the place where the electricity is used, a lot of energy will be wasted. It is important to build power stations near to industrial centres whenever possible.

Answer **D.**

24 You have to use scientific principles to decide how this device should work. Would it control the supply of gas to the boiler properly, giving less gas as the boiler heats up and more gas as the boiler cools?

(a) The volume expands when temperature increases. *(1)*
(b) Piston P moves to the left. *(1)*
(c) This starts to close hole H and reduces the flow of gas to the boiler. *(1)*
(d) What does produce this force? Is it gravity, the pressure of the gas, the compressed spring or the contraction of liquid L? There may be more than one thing causing the force. *(1)*

25 What differences can you see?

(a) Different width of fabric.
(b) Slightly different length of fabric.
(c) Different spring balances.
(d) Different number of fixings at the bottom.

Are there others? Let us look at the above differences in more detail.

(a) Suppose a 1 cm width of cloth tears with a force of 3 N. Would a 10 cm width of cloth also tear with a force of 3 N? This is unlikely. The 3 N would be shared out across the 10 cm width as 0.3 N per cm. So width is important.
(b) Fabric A is hanging loosely in the diagram. When stretched out it would be longer than fabric B. Suppose that you had two samples of fabric A, each 1 cm wide and having 20 threads in that width. If there is a force of 20 N, it is 1 N per thread in both cases. No matter what the length of the samples, the force is still 1 N per thread. But a thread usually breaks at a weak point. Is the chance of a weak point the same in a 10 cm length and a 20 cm length? Does length matter?
(c) So long as both spring balances work accurately, it does not matter that they are different.
(d) If the pulling force is 3 N, there will be a force of 3 N on just one fixing or a force of 1 N on each of three fixings. But it is cloth that is being tested, not the fixings, so it does not matter whether there is one or three.

32 The experiment is trying to find out how much different people sweat. What could this mean? Who sweats most? Who sweats least? The average rate of sweating for the eight people? The quantity of sweat produced by each person when sweating at the maximum rate? *(1)*

The experiment does not take the size, age or fitness of the students into consideration. So the results may not be typical of people in general. *(1)*
The men drank some liquid just before the experiment.

The experiment was not designed to take account of anything that the students did before the experiment began. Drinking may increase sweating by some people. Another experiment would be needed to find out about this. *(1)*

The men and women used the sauna on different nights. There may have been a difference in temperature. The number of people going in and out, opening the door, may have been different on the two nights. This could have affected conditions in the sauna. *(1)*

If Tom passed 500 g of urine when he visited the toilet, his mass at the end of the experiment would be 500 g too low. It would look as though the 500 g lost was sweat. *(1)*

The body sweats when it is surrounded by very hot air. Does it matter if half the body is wrapped in a towel or the back surfaces are in contact with a bench? What do you think? (Hints: (a) David Livingstone, the African explorer, had thick felt clothing made for his journey. (b) In hot countries, people can get prickly heat in places where tight clothing prevents evaporation of sweat. This often happens at the waistband of trousers or skirts. It is due to the skin becoming too hot when sweat cannot evaporate.) *(2)*
Does exercise alter sweating? *(1)*
What about Alan's chewing gum? *(1)*

38 You have to evaluate the arguments in terms of scientific principles.

(a) Yes, plastic bubbles do trap air and air is a bad conductor of heat. *(1)*
There is at least as much air in Harry's cavity. (The plastic itself takes up a bit of space in Abdul's.) *(1)*
The air in Harry's cavity is not trapped. *(1)*
Hot air can rise and escape. The house is cooled by a convection current of rising air. *(1)*

(b) Has Harry got his arithmetic right? Is it true that the double glazing will soon pay for itself?

(c) Has Abdul got his arithmetic right? Is each m^2 of wall as important as each m^2 of window?

41 You are being asked to evaluate a technological process. But you need to remember some science facts and principles in order to do this.

(a) The metal contracts on cooling. *(1)*

(b) It could start to solidify before pouring is complete. *(1)*

(c) To protect him from the heat given out by the melted metal. *(1)*
To protect him from burns. *(1)*

(d) Density = $\frac{\text{mass}}{\text{volume}}$

Volume = density × mass *(1)*
= 2800 × 0.002 *(1)*
= 5.6 kg *(1)*

(e) Some of the metal will remain in the container after pouring. *(1)*
Some of the metal will evaporate while it is liquid. *(1)*

5 Synthesis

To deal with real situations in science, you need to be able to use all the science skills at the same time. The questions in this section involve at least two of the skills which were dealt with in the earlier part of the book. To answer some of them you will be handling information, solving problems, using patterns, evaluating and applying knowledge of science and scientific principles all in the same question. Bringing all the pieces together to make something is called synthesis.

Multiple choice questions usually test only one thing. So there cannot be any synthesis multiple choice items. To give opportunity for you to use all the skills, some of the questions are quite long.

Some GCSE science examinations will have questions with at least twenty-five marking points which will take you at least twenty minutes. So these longer questions may be good practice for GCSE.

2 This question begins with information handling. Later you have to solve a problem.

(a) From the paragraph: the lower the *U*-value, the lower the loss of heat. So to answer this question, you want the glazing which has the highest *U*-value. This is single glazing with metal frames. *(1)*

(b) If better means losing less heat, (rather than cheapest or most attractive) you want to know which has the lowest *U*-value. Yes, it is 2.5 for wooden and 3.2 with metal frames. *(1)*

(c) The fully insulated roof has a *U*-value of 0. It loses no heat. So insulate the roof fully. *(1)*

(d) This is the roof with no insulation. *(1)*

(e) A low *U*-value means low heat loss. *(1)*
It means that the material is a good insulator. It may mean that the material is a poor conductor (wood rather than metal). It may mean that the material is quite thick. 'What does a low *U*-value mean?' is an 'open-ended' question. There could be more than one answer.

In an easy paper the first sentence should be sufficient to get full marks. In a hard paper (for grades A and B), more information might be expected.

(f) Formula
$T = 40 - 30 = 10°\text{C}$ *(1)*
$H = A \times T \times U$
$= 100 \times 10 \times 1.5$ *(1)*
$= 1500$ W (or 1.5 kW) *(1)*

4 Information is presented as maps. Symbols are used to represent different plants. Take care not to confuse moss and grass. You are comparing the patterns. There are similarities and differences. You have to pick out pieces of information. You also have to draw conclusions about what the information might mean.

(a) Lichen is on the stone in September but in March all of it has gone. It looks as though it could have died. *(1)*
(b) Grass covers less area in March than in September. The part which has gone is over gravel. *(1)*
(c) Snowdrops. *(1)*
(d) What produced the dead leaves? *(1)*
(e) Gravel is usually dry. Moss grows in damp places. Water flows downhill. Is gravel higher up than the moss?
The dead leaves are on the moss. They would tend to fall to the bottom of a slope. *(1)*

6 You are given ten pieces of information. This must be organized into a pattern. You will have to make judgements about which decay chain the information fits into and how it links up with other parts of the chain.

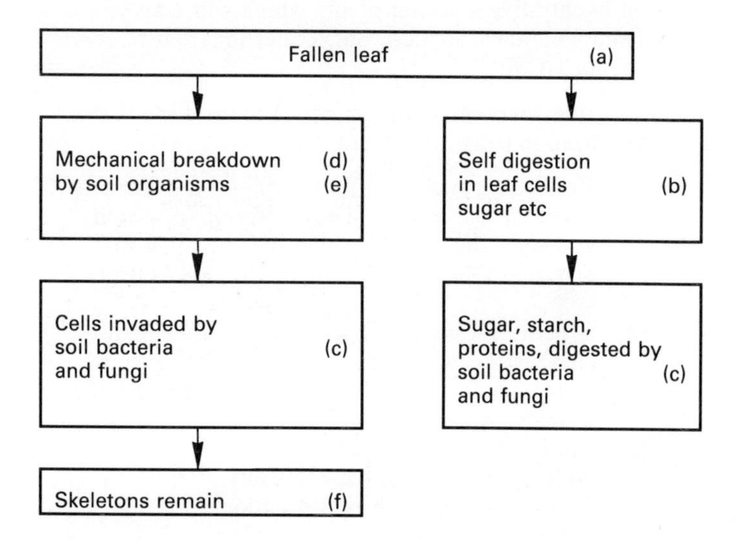

Can you complete it?

Marks

Separation of leaf structure change from molecular change.
Complete *(2)*
Partial *(1)*

Start at (a) *(1)*
Finish at (j) *(1)*

Sequencing
Chart 1
Complete *(2)*
Partial *(1)*
Confused *(0)*

Sequencing
Chart 2
Complete *(2)*
Partial *(1)*
Confused *(0)*

Parallels indicated
(c) – (c) *(1)*
(g) – (g) *(1)*

15 (a) Take care with the scale. It is a 'logarithmic scale'. There are only nine divisions between 0.1 and 1. They get smaller in size as you go towards 1. You must take account of the uneven scale when you are getting the longest wavelength of red light. It is between 0.70 μm and 0.80 μm. But what is the exact wavelength? The middle of the spectrum must be halfway between the ends. So you must calculate the average of the wavelengths for (i) and (ii). *(3)*

(b) (i) You saw in diagram 1 that there is ultraviolet radiation with wavelengths less than 0.1 μm. For wavelengths less than this, platinum seems to be the best. *(1)*
(ii) The curves run close together so you must take care. Gold seems to be best for wavelengths above 3 μm. *(1)*
(iii) Now you have to bring in your answer for (a)(iii) and use it to get a reading from graph 1. Even if (a)(iii) was wrong, you would get credit for using that answer correctly here. *(1)*
(iv) You have to make an evaluation of the data. *(1)*

(c) Ultraviolet radiation lies between visible light and X-rays. Could X-rays pass through you? Could they pass through a thin layer of aluminium? What does transparent mean? *(1)*
What happens to the reflected light? *(1)*

(d) Three of them are precious metals. They are valued because they retain their polished surface. Gold and platinum do not deteriorate in air. Silver deteriorates more slowly than most metals. Aluminium is self protecting. What is the pattern? *(1)*

(e) (i) The atmosphere absorbs some radiation. *(1)*
All radiation of wavelengths less than 0.3 μm is absorbed. *(1)*
The amount of absorption between 0.3 μm and 0.4 μm is variable
OR
The percentage of radiation absorbed is greater for shorter wavelengths. *(1)*
(ii) How can the problem be solved?
Above the atmosphere or high in the atmosphere. *(1)*
In a satellite or on a mountain top. *(1)*

(f) You are evaluating these statements. Do they fit in with scientific principles?
(i) Reflection from snow, *(1)*
upwards! *(1)*
(ii) High in atmosphere, so very much more ultraviolet. What is the most effective wavelength: graph 3? What does the atmosphere do at this wavelength: graph 2? *(1)*

Reflection from snow. *(1)*
(iii) Less oxygen per lungful of air *(1)*
Breathing equipment covers the nose! *(1)*

21 (a) This is a problem.
48 – 12 = 36 years *(1)*
28 days = 4 weeks *(1)*
There are $\frac{52}{4}$ = 13 ovulations each year *(1)*
36 × 13 = 468 eggs *(1)*
You may know that a woman does not ovulate or menstruate during pregnancies and this may continue for a few months following the birth. This is not stated in the question. It has not been taken account of in the answer. How would the answer be altered if the woman had four 9 month pregnancies between the ages of 12 and 48?

(b) You have to link 'egg released' in diagram 2 with 'ovulation'. How do the hormone levels change? *(1)*

(c) (i) Look at the arrow in diagram 2 going from the egg to the wall of the womb. Compare this with the bottom axis showing time in days. *(1)*
(ii) You need to do some arithmetic to solve this problem. Sperm in the woman two days before ovulation could produce a pregnancy. Once the egg has reached the wall of the ovary there is no chance of pregnancy. So the answer is up to two days before ovulation and . . . . *(1)*

(d) (i) At about 36.9°C from the start of menstruation until just before ovulation. *(1)*
At about 37.7°C from ovulation to menstruation. *(1)*
At ovulation a brief temperature fall (followed by the temperature rise). *(1)*
(ii) Illness, such as a sore throat. *(1)*

(e) (i) This is a problem-solving exercise. Use the thermometer to detect the start of ovulation. How? *(1)*
Ensure that intercourse takes place soon after. *(1)*
(ii) This calls for evaluation. How would sodium hydrogen carbonate affect the acid? Does acid encourage the sperm to go from the vagina to the womb? What about the survival of the sperm? You should produce an argument containing at least three reasonable ideas. *(3)*
The position of fluid is affected by gravity. *(1)*
What position would be least likely to allow the sperm to reach the egg tube? (Position is important after as well as during intercourse.) *(1)*
(iii) Sweating causes cooling. *(1)*
In what conditions will sweat be produced? *(1)*
In what conditions will the muscles contract to hold the testicles against the body? *(1)*
Clothing should not hold the testicles against the body, for example, tight jeans. *(1)*
and should allow sweat to evaporate, for example, cotton underwear. *(1)*
(Sperm take about six weeks to develop. Sperm development might stop while a man is taking a hot bath. But problems are only likely to occur if the automatic cooling mechanism is prevented from operating for long periods of time.)

Acknowledgements

The authors are grateful for permission to use data which has appeared in journals of the Institute of Biology and mode 3 examinations of the Yorkshire and Humberside Regional Examinations Board.

Thanks are also due to Julia Bennetts, who typed the manuscript of this book.

Prefixes used in this book

Symbol	Prefix	Meaning	Example	
μ	micro	$\frac{1}{1\,000\,000}$	μm	(micrometre)
m	milli	$\frac{1}{1\,000}$	mm	(millimetre)
c	centi	$\frac{1}{100}$	cm	(centimetre)
d	deci	$\frac{1}{10}$	dm	(decimetre)
k	kilo	1000	km	(kilometre)
M	mega	1 000 000	MJ	(megajoule)

British Library Cataloguing in Publication Data
Science skills problems in GCSE science.
1. Science — Questions & answers — For schools
I. Bennetts, J.
507'.6

ISBN 0 340 41485 5

First published 1988
Second impression 1989

Printed and bound in Great Britain by
Courier International Ltd, Tiptree, Essex
for Hodder and Stoughton Educational
a division of Hodder and Stoughton Ltd, Mill Road
Dunton Green, Sevenoaks, Kent